Number Stories

Learning Arithmetic Through the Adventures of Ralph and His Schoolmates

By Alhambra G. Deming

Principal of the Washington School, Winona, Minnesota

PANTIANOS
CLASSICS

Published by Pantianos Classics

ISBN-13: 978-1-78987-095-4

First published in 1916

Contents

Preface

Under the social ideal we take problems which have to do with the child's life, with his world, and by leading him to solve these problems, the problems in his own world, we hope to create a habit of mind that will help him to solve the problems in the world that he is to know later.

- Henry Turner Bailey

These stories are to be read to pupils in the intermediate grades. Their primary aim is drill in the essentials of arithmetic as applied to child experience.

In addition to this the vital lessons of system, industry, independence, uprightness, courtesy, school loyalty, generosity, thrift, and appreciation and consideration of parents are taught by suggestion.

The *story* should be kept uppermost and the teachings, apparently, made incidental.

Alhambra G. Deming

Introduction

Arithmetic as taught in the past has, perhaps, been more abstract and more foreign to the real life of the child than any other subject included in our grammar-school curriculum. The pupil's imagination might help compass the end in view in the teaching of geography, reading, and history, but in arithmetic the imagination has been offered no opportunity to help. Since school-teaching began children have juggled with the uninteresting symbols which we call figures, without realizing their values and without a suspicion that they could have anything to do with a child's experience. Even when the abstract was made concrete by the application of figures to actual working experience, the conditions of the problems presented were generally far removed from the child's interest. Such problems were for the most part made to fit grown-up conditions, and were utterly dissociated from one another.

The past few years have seen much improvement in the teaching of arithmetic, but there is still crying need for methods which bring this subject into closer relationship with the life of the child. These Number Stories have been written with the object of begetting a live interest in a hitherto abstract subject.

Interest begets effort; effort begets accuracy; accuracy begets efficiency, and efficiency is to-day the great aim of elementary education.

In the hope that these stories may contribute somewhat toward this great end, I hereby dedicate them to those who lay the foundation upon which the High School and College build — the Intermediate Grade Teachers.

Problems Presented

HOW RALPH HELPED

Saturday

1 Amount of money in Ralph's bank

2 Amount taken out

3 Amount put back

Monday

4 Number of papers bought by Ralph

5 Amount received for them

6 Amount really made

7 Balance in bank at night

Tuesday

8 Amount given Ralph for returning purse

9 Balance in bank at noon

10 Amount made on papers

11 Balance in bank at night

Wednesday

12 Amount received for rags

13 Amount made on papers

14 Balance in bank at night

Thursday

15 Amount earned by mowing lawn

16 Weight of magazines given by Mrs. Baxter

17 Amount paid for them by junkman

18 Amount made on papers

19 Total earnings

20 Balance in bank at night

Friday

21 Number of nickels received for best bananas

22 Amount received for them

23 Amount received for second-grade bananas

24 Number of bananas sold to workmen

25 Amount received for them

26 Total amount received for bananas

27 Ralph's share

28 Amount made on papers

29 Ralph's total earnings

30 Balance in bank at night

Saturday

31 Cost of lemons bought by Ralph

32 Cost of ice and sugar

33 Total cost of lemonade

34 Amount received for it

35 Amount cleared

36 Balance in bank at noon

37 What Ralph thought he had earned

38 Amount really earned

39 Amount paid him by Uncle Ben

40 What Ralph gave his mother for the doctor

41 His mistake

RALPH'S SUMMER ON THE FARM

I

1 Number of miles from Ralph's home to Grandpa's farm

2 Number of hours required for trip

3 Cost of trip

4 Amount of change given Ralph by Uncle Ben

II

6

5 Amount left if entire trip had been by train

6 Number of years spent by Grandpa on farm

7 Grandpa's age

8 Grandpa's flax crop

9 Value of crop

III

10 Number of acres in farm

11 Plan of farm (dictation)'

12 Number of miles of fence around farm

13 Cost of fence

IV

14 Number of chicks hatched by Grandma 's hens

15 Number of eggs "wasted" in hatching

16 Number of head of stock owned by Grandpa

17 Value of wool sheared

18 Value of sheep sold

19 Value of sheep and wool sold

V

20 Amount earned by Jerry in a year

21 Total amount earned by Jerry

VI

22 Distance ridden by George with Ralph and Grandpa

23 Amount received by Grandpa for butter, eggs, and strawberries

24 Cost of Grandpa's groceries (bill)

25 Amount of money taken home

VII

26 Number of trees required for telephone poles, from farm to town

27 Number of apple trees in orchard

28 Number of bushels of apples expected

29 Number of barrels

30 Estimated value of apple crop

VIII

31 Number of bushels of oats expected

32 Number of bushels to be sold

33 Amount expected for oats

34 Number of rods of fence around farm

35 Number of posts in fence

IX

36 Value of cherry crop

37 Value of last year's plum crop (three ways)

X

38 Estimated potato crop

39 Number of bushels to be sold

40 Value of potatoes sold

41 Estimated value of wheat crop

42 Estimated value of com crop

43 Estimated value of buckwheat crop

44 Estimated value of hay crop

XI

45 Probable value of pumpkin crop

XII

46 Number of apples earned by weeding

7

47 Number of apples earned by picking cherries

48 Total number of apples earned

49 Number of eggs laid by hens in two months

50 Number of dozen laid in two months

51 Total value of eggs

52 Amount earned by each hen

XIII

53 Amount expected by Grandpa for his entire crop

54 Number of nights spent by Ralph at farm

55 Number of meals eaten by him there

56 Cost of meals

HOW EVERYBODY HELPED

Amount Contributed for Radiopticon

HUSTLERS

1 Floyd Aldrich
3 Jean Bryce
5 Walter Brown
7 Ema Lawton
9 Chris Altman
11 David Winchester
13 Margaret Whitten
15 Ernest Belden
17 Sallie MacKay
19 Peter Lewis
21 Jimmie O 'Hagan
23 Gertrude Blake
25 Jerome Sanders
27 Louise Farrel
29 Ralph Merton

RUSTLERS

2 Dennis Ryan
4 Genevieve Hart
6 Carl Cummings
8 Sigmund Wolski
10 Eva Moreland
12 Joe Beverley
14 Kate Caswell
16 Leonard Monroe
18 Hans Schmidt
20 Annabel Jones
22 Dorothy Perkins
24 Fay Leland
26 Effie Wheaton
28 Mabel Gartney
30 Flora Aldrich

===

31 Total contribution in Sixth Grade (three processes)

32 Average number of pennies given by Kindergarten children

33 Contributions of parents to First and Second Grades

34 Amount brought by pupils of Third and Fourth Grades

35 Contributions of pupils in Fifth Grade

36 Amount brought by Seventh Grade

37 Amount contributed by Eighth Grade

38 Total contribution of all grades

39 Total after adding teachers ' contribution

40 Balance after deducting cost of radiopticon

41 Amount made at candy sale

42 Total from Eighth Grade after candy sale

43 Amount of victrola fund after candy sale

44 Amount made at radiopticon entertainment

45 Probable attendance

46 Entire victrola fund

THEIR COST

1 Number of meals eaten per year

2 Number of meals eaten in entire life

3 Entire cost of Ralph's meals

II

4 Cost of estimated clothing for a year (twelve processes)

5 Cost of estimated clothing for entire life (twelve processes)

III

6 Cost of benevolences, (a) for one year; (b) for entire life

7 Cost of amusements, (a) for one year; (b) for entire life

8 Cost of treats, (a) for one year; (b) for entire life

9 Cost of travel, (a) for one year; (b) for entire life

10 Cost of illness, (a) for one year; (b) for entire life

11 Cost of school material, (a) for one year; (b) for entire life

12 Total cost of each pupil for entire life

13 Total cost of Ralph Merton for entire life (involving thirty-five processes)

14 Dorothy Perkins's cost (thirty-five processes)

HOW THE CITY HELPS

1 Amount of Mr. Brooks 's tax

2 Amount of Mr. Merton 's tax

3 Amount of Mr. Perkins 's tax

4 Amount of Mr. Brooks 's school tax

5 Amount of Mr. Perkins 's school tax

6 Amount of Mr. Farrel's tax

7 Mr. Farrel's city school tax

II

8 Annual cost of writing supplies (bill)

9 Annual cost of drawing and handwork supplies

III

10 Annual cost of supplies for janitor

11 Annual cost of miscellaneous supplies

12 Total annual cost of supplies (See Chapter II)

IV

13 Cost of readers (seven problems)

V

14 Cost of arithmetics and spellers (three problems)

15 Cost of music readers, geographies, and hygienes (four problems)

VI

16 Cost of geographical readers, history readers, and special books (three problems)

17 Total cost of textbooks, without readers (See Chapter V)

VII

18 Cost of reference books, maps, globes, and charts (two problems)

19 Total cost of books in building

20 Average cost of books for one year

VIII

21 Cost of drawing and handwork appliances

22 Cost of janitor's appliances

IX

23 Cost of domestic science equipment

24 Cost of manual training equipment

25 Cost of miscellaneous appliances

26 Cost of orchestra equipment

27 Cost of playground equipment

28 Total cost of appliances and equipment

29 Cost of school furnishings

30 Cost of trees and shrubbery

31 Cost of permanent school property

32 Total annual salary of special teachers and officials

33 Our share of annual salary of special teachers and officials

34 Total of annual salaries used for our benefit

35 Amount of other annual running-expenses

36 Total amount of annual running-expenses

37 Interest on value of permanent property for one year

X

38 Virtual amount expended annually

39 Average amount expended annually for each pupil

How Ralph Helped

Suggestions

Supply each pupil with several sheets of paper at least 5 in. x 8 in. On one sheet let him write the name of the story, and his own name. This will serve as the cover of a booklet in which will be placed the results of Ralph's money-making.

At the top of the first page write, "Saturday," and under this write the problems belonging to that day. Use a separate page for each day.

In lower grades take the experiences of one day, or even part of a day, at a reading, adapting them to conditions.

In higher grades, this story and those that follow are excellent for teaching rapidity and accuracy in reckoning. Here the work should be taken mentally, as far as possible, amounts for future reference being recorded on the blackboard instead of in the booklets.

The End of School

It was Friday, the twelfth of June, and school had closed. How happy the children were! — now they could play from morning till night! Ralph Merton was one of the happiest of them all, for he had been promised a visit to Grandpa's farm. There he could gather eggs, feed the chickens, ride the horses to water, and do all sorts of other interesting things.

But Saturday morning something dreadful happened in the Merton family. Ralph's father, who was a carpenter, fell from a roof on which he was working, and was badly hurt. When the doctor came, he looked very grave.

"It's a wonder you are alive, after such a fall!" he said. "However, I think we shall bring you out all right. But you must lie here quietly for several weeks."

"What will my wife and boy do!" moaned poor Mr. Merton. "Oh, I *must* work!"

"You see," Mrs. Merton explained to the doctor, "we have just paid for our house, and we haven't a penny saved to live on. But don't worry, James," she said to her husband; "we'll manage, somehow. I can do sewing and other kinds of work till you are well. Don't worry, dear."

Ralph was only eleven years old, but he was a manly, thoughtful boy. He listened to the conversation and then went out and sat down on the back steps to think it over. He then and there resolved that he, too, would help, so that his father need not worry,

"Then he'll get well more quickly," he said to himself; and he bravely add-ed: "Good-by, farm; you won't see me this summer!"

He went into his room, opened his little bank, and laid his money out on the bed. There were the big round dollar Uncle Ben had sent him for Christ-mas; the fifty-cent piece his father had given him for bringing in the wood every day in January and February without once being reminded; and the silver quarter his mother had slipped into his hand one day when, on looking at his report card, she found that his mark in Number was ninety-five. (Num-ber was his hardest subject.) Besides, there were three nickels, two ten-cent pieces and twenty-nine shining pennies — for when his father gave him pen-nies they were always new ones. He counted his money and added it up, and found he had —

How much, children?

$1.00 + $0.50 + $0.25 + $0.15 + $0.20 + $0.29 = $2.39, *amount in bank.*

However, he knew that was not nearly enough for the family to live on un-til his father got well; so he began to try to think of a way to earn some more.

He thought hard, and finally a bright idea came to him.

He had often seen small boys — smaller than he — peddling papers on the street.

"I'll take some of this money and buy some papers to sell," said he to him-self.

He took out one quarter, one nickel and sixteen pennies, which he put into his purse.

How much did he take out?

$0.25 +$0.05 +$0.16 = $0.46, *amount taken out.*

The remainder he put back into his bank. How much did he put back?

$2.39 -$0.46 = $1.93, *amount in bank Saturday night.*

Monday

On Monday morning Ralph ran several errands and then sat with his fa-ther while his mother finished her housework. After the midday dinner he went to a news stand where he had seen other boys buy papers.

"Can I buy some papers here, to sell?" he asked the dealer.

"Certainly, "replied the man, "if you have money to pay for them."

"How much are they? "asked Ralph.

"Two cents apiece for this kind," the news dealer answered.

"I have forty-six cents," said Ralph, taking out his purse. "Give me —"

How many papers could Ralph buy?

$0.46 ÷ $0.02 =23 *times.* 23 *papers.*

"How much shall I charge for them?" he asked.

"Sell them for five cents apiece," said the man.

Ralph went about the streets calling out his papers as he had heard the other boys do. Sales went slowly at first and he was growing discouraged, when he saw a group of men standing on the hotel steps. As they saw him

12

approaching, one man said to the others:

"There's that Merton boy, selling papers. His father was almost killed last Saturday — fell from a roof. He's as manly a boy as I have ever seen. Let's buy him out.

"Hi, there, youngster!" he called to Ralph, "bring your papers here!"

The men on the steps bought every one of Ralph's remaining papers. When he counted the money in his purse, he found he had —

How much?

$23 \times \$0.05 = \$1.15$, *amount received for papers.*

"But I didn't really make all of that," said he to himself, "for they cost me forty-six cents. I have made only —

How much had he made?

$\$1.15 - \$0.46 = \$0.69$, *amount really made.*

Of course, when Ralph reached home his mother asked where he had been so long, but he said:

"Please, Mother, don 't make me tell now! I've planned a surprise for you, and I promise to tell you all about it next Saturday."

"Very well, Ralph," she replied. "I'm sure I can trust you."

He ran into his room, opened his bank, and put in his newly earned money.

"Now," said he, "I have —

How much had he?

$\$1.93 + \$1.15 = \$3.08$, *amount in bank Monday night.*

Tuesday

The next morning Ralph's mother sent him to the post office with a letter.

"Be careful not to lose it," she said. "It is to Uncle Ben, telling him of your father's accident. "

"Yes, Mother," Ralph replied, and with a hop, skip, and jump he went down the street, and into the post office. He dropped the letter into the box, and turned toward the door. Just ahead of him was a tall man who had been buying stamps and who had his hands full of mail. As he passed through the door he dropped a big, fat purse, but he walked oil without noticing his loss. Ralph picked it up, and looked at it wistfully.

"Very likely there's enough money in it to keep us until father gets well," thought he. "But it isn't mine," he added quickly, and he ran after the man, calling: "Mister, you dropped your pocket book."

The man turned, and Ralph saw a very kind face, lit by a friendly smile.

"How is it that you didn't keep?" asked the man quizzically.

"Because it isn't mine," answered Ralph. "I shouldn't want any one to keep my purse if I lost it."

"And a very good reason," said the man. "Here, hold out your hands." He opened the purse and emptied into Ralph's outstretched hands all the loose change it contained. "Take that for your trouble."

"Oh, thank you, sir!" cried Ralph. "I wouldn't take it except that my father has been hurt and we need it so much," he went on; but the man was already hurrying away.

Ralph looked at the money still in his hands. Then he sat down on the curb stone and counted it. There were five quarters, seven nickels, six dimes, two fifty-cent pieces, and twelve pennies. In all there was —

How much?

2 x \$0.50 =\$1.00, *value of fifty-cent pieces.*
5 x \$0.25 =\$1.25, *value of quarters.*
6 x \$0.10 =\$0.60, *value of dimes.*
7 x \$0.05 =\$0.35, *value of nickels.*
12 x \$0.01 =\$0.12, *value of pennies.*
———
\$3.32, *amount received.*

"That man must be rich, to give away money like that!" exclaimed Ralph.

When he rushed into the kitchen, where his mother was finishing her morning's work, he looked so excited that she asked:

"Why, Ralph, what is the matter?"

"Something wonderful has happened," he answered. "But it's a part of the surprise, and I can't tell you about it yet," he added, and he laughed as he ran into his room to put away his new money.

"Let me see," said he, "I had three dollars and eight cents before and that kind man gave me three dollars and thirty-two cents. Why, I'm almost as rich as he! I have —

How much had he in all?

\$3.08 + \$3.32 = \$6.40, *amount in bank Tuesday noon.*

Every afternoon, at about the time the men were going home from work, Ralph sold papers. When the news dealer saw how well he had done on Monday, he told Ralph he might take twenty papers each day, and return those which he could not sell. That day he sold twelve, on which he made —

How much?

12 x \$0.03 = \$0.36, *amount made on papers.*

He added this to his bank money; so on Tuesday night he had in all —

How much?

\$6.40 +\$0.36 =\$6.76, *amount in bank on Tuesday night.*

Wednesday

"Mother, may I sell the bag of rags you have up in the attic? asked Ralph the next morning.

"Yes, if you want to," replied his mother, "and I wish you would take away those pieces of old carpet that are lying in the shed. They are so unsanitary. "

Ralph brought down the bag of rags and stuffed them and the old carpet into a big sack. Loading this upon his express wagon, he hauled it to the junk-dealer's.

"How much you got?" asked the junkman. "I haven't weighed it," answered Ralph.

The man put the sack on the scales and found it weighed just twenty-six pounds.

"I pay you one and half cent a pound," said he, and handed Ralph the money.

How much did he hand him?

26 x $0.015 = $0.39, *amount received for rags.*

As Ralph turned to leave, the junkman said:

"Say, maybe you got some magazines you want to sell. I give you half-cent a pound for old magazines."

That night Ralph put his thirty-nine cents — "rag money," as he called it — into his bank. He had sold eighteen papers that afternoon, on which he had made —

How much?

18 x $0.03 =$0.54, *amount made on papers.*

So when he counted his money just before going to bed he found he had — How much in all?

$6.76 + $0.39 + $0.54 = $7.69, *amount in bank Wednesday night.*

Thursday

Mrs. Baxter lived in a big white house, which had a beautiful green lawn in front of it. Ralph's mother had sent him to the house with a note, asking if Mrs. Baxter could give her some plain sewing to do. While Ralph was waiting for a reply he had time to notice that the grass on the lawn needed to be cut

Presently Mrs. Baxter came out of the house with a bundle of sewing for his mother.

"Don't you want to hire a boy to cut your grass, Mrs. Baxter?" he asked. "I'd like the job."

"I shall be glad to have you do it if you are strong enough," Mrs. Baxter replied. "The man who usually does it has been sick for the past few days, and I don't know when he can come again. I pay him twenty cents an hour, and I will pay you the same."

"Thank you," said Ralph, "I'll be back in half an hour."

He took the bundle of sewing that Mrs. Baxter held out and hurried home.

As he gave the bundle to his mother, he asked, "Mother, may I go somewhere until dinner-time?"

Receiving permission, he hurried back to the big house and was soon at work mowing the lawn. At first it was great fun, but as the morning grew warmer the lawn mower grew heavier, and he finally said to himself:

"This is too much like work! I believe I'll give it up!"

But when he thought of his sick father, and how much bigger this work would make his "surprise," he kept bravely on. Of course he rested once in a while, but even a grown man does that, and nobody counts it out of his time. At the end of three hours and a half he had finished his work.

How much did he earn?

3 ½ x $0.20 =$0.70, *amount earned.*

When Mrs. Baxter paid him his money, and then gave him a large glass of milk, how glad he was that he had not given up!

"By the way, I have some old magazines in my attic," said Mrs. Baxter. "If you will carry them down to the basement, I will pay you fifteen cents. You will have to make several trips, for there is a big pile of them," she added.

"What do you do with your old magazines!" asked Ralph, remembering what the junkman had told him.

"I bum them in the furnace," she answered.

"As you don't use them, will you give them to me if I carry them down for nothing!" asked Ralph.

"Yes, indeed! I shall be glad to, for they do not bum easily," said Mrs. Baxter.

Magazines are heavy, and the attic was hot, but Ralph persevered until he had carried eight armfuls down the two flights of stairs, and made eight neat piles of them on the back porch.

"May I leave them until after dinner?" asked he. "Then I'll bring some string and tie them up."

After dinner he tied the magazines up into bundles, packed them into his express wagon, and took them to the junkman.

"Will you weigh each package separately, so I can see how many pounds I carried downstairs each time?" asked Ralph.

The junkman good-naturedly did as Ralph requested, and they found the packages weighed, respectively, twelve and one-half pounds, eleven and one-half pounds, ten and one-half pounds, nine and one-half pounds, fourteen pounds, thirteen pounds, eleven pounds, and twelve pounds. Then the man weighed them all together and they weighed—

How much in all?

12 ½ + 11 ½ + 10 ½ + 9 ½ + 14 + 13 + 12 = 94. – 94*lb*

The junkman paid, him one-half cent a pound. How much did he receive f

94 x $0.005 = $0.47, *amount received for magazines.*

That afternoon Ralph sold thirteen papers. That brought him in just —

How much!

13 x $0.03 = $0.39, *amount made on papers.*

"That is just equal to my rag-money," said he. "Let me see, how much have I made to-day? I earned seventy cents by mowing the lawn, I sold the magazines for forty-seven cents, and I made thirty-nine cents on papers. That's —

How much was it?

$0.70 +$0.47 +$0.39 =$1.56, *total earnings.*

"That is the best I've done any day, "said Ralph. He added it to his bank fund and found he now had —

How much in all?

$7.69 +$1.56 =$9.25, *amount in bank Thursday night.*

"I'm getting rich fast, "said he.

Friday

"My boy," said Mr. Barnes, the grocer, as Ralph stood waiting for the groceries that were being put up for him, "I hear you 're getting to be a real man of business. Do you want another job?"

"I should say I do!" Ralph answered earnestly. "I'm looking for jobs, these days."

"Do you see that pile of bananas over there? They are getting so ripe that I am afraid they are going to spoil, and I have a lot more coming. If you will peddle them out, I will give you half the money you take in. You can sell the best ones at three for five cents, the next grade for a cent apiece, and the rest for whatever you can get. Do you want to try your luck?" asked the grocer.

"Indeed I do!" exclaimed Ralph. "I will take these groceries home, and come right back with my express wagon."

His mother said she did not need him until dinnertime, so he hurried back to the grocery store. He sorted the bananas carefully, put them into his wagon, and went to the kitchen doors of all the people he knew and some he did not know.

"They are very ripe," he said, "and they won't keep long, but you might use them for dinner."

He was very polite, and almost everybody knew about his father 's accident, so it was not long before he had sold thirty-six of the very best ones — the three-for-a-nickel kind.

How many nickels did he get?

$36 \div 3 = 12$ *times.* 12 *nickels received for best bananas.*

How much money did he receive? 12 x $0.05 = $0.60, *amount received.*

He sold forty-five of the penny-apiece ones, which of course was just —

How much?

45 x $0.01 = $0.45, *amount received for second-grade bananas.*

He now had left only the poor ones, which he had almost given up hope of selling, when he saw a large group of men who had been working on the street, just sitting down in the shade to eat their lunch. Ralph walked over to them.

"Don't you want to buy some bananas to eat with your lunch?" he asked. "They are very ripe, but if you eat them right away they will be all right, and they are only eight cents a dozen."

"Let's look at 'em, sonny," said one of the workmen, and then he selected one dozen. Another man took a dozen and a half; another three-quarters of a

17

dozen, another a dozen and a quarter, and still another a half-dozen. They passed the bananas around and most of them were eaten "right away," as Ralph had advised, while the remainder were put away in the lunch-boxes.

How many bananas did Ralph sell to the workmen? How many dozen?

12+18+9+15+6=60, *number of bananas sold.*

60 bananas ÷ 12 bananas = 5 *times. 5 doz.*

How much did he receive for them?

5 x $0.08 = $0.40, *amount received for bananas.*

With his empty wagon rattling behind him, Ralph hurried back to the grocery store, and so excited was he that he fairly shouted:

"Mr. Barnes, I've sold every banana! I've sold out!"

"Well, you are a good salesman!" responded Mr. Barnes heartily. "Come in and we will settle up."

Then Ralph showed his money and told how he had received sixty cents for the best bananas, forty-five cents for the next best, and forty cents for the very ripest, which amounted to —

How much?

$0.60 +$0.45 +$0.40 =$1.45, *total amount received for bananas.*

"What is your share, boy?" asked the grocer, just to see whether or not Ralph knew.

What was Ralph's share?

$1.45 -2 =$0,725, *Ralph's share.*

"Seventy-two and a half cents, Mr. Barnes," he answered promptly; for he had already thought it out.

"Good for you, boy! I will give you the extra half-cent, which makes your share exactly —

How much did Ralph receive?

$0.725 +$0.005 =$0.73 *paid Ralph.*

Ralph went miming home to dinner, the hungriest and happiest boy in town.

That afternoon he sold eighteen papers, which amounted to —

How much?

18 x $0.03 = $0.54, *amount made on papers.*

At night, when he counted his money, he found he had earned seventy-three cents and fifty-four cents that day, which is —

How much!

$0.73 +$0.54 = $1.27, *amount earned on Friday.*

When he added this to the nine dollars and twenty-five cents in his bank, he had in all —

How much?

$9.25 + $1.27 =$10.52, *amount in bank Friday night.*

"I'm almost a millionaire, "said Ralph as he cuddled down in bed. "Oh, won't mother be surprised!"

Saturday

"Circus-day!" exclaimed Ralph, as he opened his eyes bright and early Saturday morning. "I know how I can make some more money to-day!" and he jumped out of bed and was bathed, dressed, and combed almost before one could have said "Jack Robinson!"

He hurried to do his regular "chores" and then ran down to the grocery store.

"How much are lemons, Mr. Barnes?" he asked.

"They are thirty-five cents a dozen," answered the grocer.

"I will take two dozen," said Ralph.

"You must be going to make some lemonade," said Mr. Barnes.

"Yes," replied Ralph, "I thought I could sell some, as this is circus-day. But the trouble is I don't know just how to make it, and mother is so busy. Besides, I wanted to surprise her," he added, "because we need money so badly, now that my father is hurt."

"See here, boy," said the sympathetic grocer, "I know how to make lemonade. How would it do to put that old counter out on the edge of the sidewalk, and then make lemonade in those big crocks? They are here to be sold, but lemonade won't hurt them. You can use these glasses, too." He pointed to a shelf where various kinds of dishes were piled up.

"Good!" cried Ralph. "Oh, thank you so much, Mr. Barnes!"

So the grocer helped Ralph move the counter to the sidewalk. They washed the crocks and glasses, squeezed the lemons, added the sugar and water and a piece of ice, and by the time people began to gather for the circus parade, Ralph was ready for business. He had used two dozen lemons at thirty-five cents a dozen, which amounted to —

How much?

2 x $0.35 = $0.70, *cost of lemons.*

Besides, he had used two and a half pounds of sugar at ten cents a pound, and five cents * worth of ice, which was —

How much?

2 ½ x $0.10 = $0.25 + $0.05 = $0.30, *cost of ice and sugar.*

Counting in the cost of the lemons, the entire cost of the lemonade was — How much?

$0.70 +$0.30 =$1.00, *cost of lemonade.*

The day was hot and the crowd had to wait so long for the parade — as the crowd always does — that the people became very thirsty. When they saw two or three buy lemonade at Ralph's stand a good many more wanted a drink — as people always do when they see others drinking — and by the time the parade came Ralph had sold sixty-two glassfuls, all he had made. The last few glassfuls were not very strong, but his customers did not seem to mind.

After the parade was out of sight, Ralph went into the store and counted his money. He knew sixty-two glassfuls at five cents each ought to be —

How much?

62 x $0.05 =$3.10, *amount received for lemonade.*

But he counted it several times before it came out that way, for there were so many pennies and nickels that he got mixed up in the counting. He then paid Mr. Barnes the dollar he owed him and found he had actually made —

How much?

$3.10 -$1.00 =$2.10, *amount made on lemonade.*

He helped Mr. Barnes put the counter, crocks, and glasses back into place, thanked him for his help, and hurried home. While his mother was preparing dinner he went into his room to put away his earnings. When he added his lemonade money" to what he had had before he found he had —

How much?

$10.52 +$2.10 =$12.62, *amount in bank Saturday noon.*

"I didn't think I could earn so much in a week," he said. "But I didn't really earn it all, because I had two dollars and thirty-nine cents in my bank to start with. "He worked an example on paper, and said, "Anyway, I have really earned —

How much?

$12.62 -$2.39 =$10.23, *amount he believed he had earned.*

He thought a moment. "No, that isn't right, either, because I didn't earn that three dollars and thirty-two cents the man gave me." He worked another example, and found that he had actually earned —

How much?

$10.23 -$3.32 =$6.91, *amount actually earned.*

"That is almost seven dollars, anyway, and I have never earned so much before. Maybe my money will pay the doctor; and what a big surprise it will be for Mother!"

He was putting his treasure safely away, when his mother called:

"Ralph, dear, go to the door, please. Somebody is knocking and I can't leave the dinner."

As he opened the door, Mrs. Merton heard a joyful shout of, "Goody! Goody! Oh, Mother, here are Uncle Ben and Aunt Lucy!"

Sure enough, there they were, with a big suit-case. When the first greetings were over, Uncle Ben explained:

"When that letter came, telling me about your trouble, we made up our minds to come right on and stay until James was better. It is too much for you and Ralph to bear alone. I can help take care of James and keep him cheered up, and Lucy can help you with the housework, and we shall get along famously."

After dinner Ralph took his uncle aside and told him about the big surprise he had been working up all the week.

"You can't imagine how much money I have earned. Uncle Ben," said he. "And every cent of it is for the doctor and the grocer, so Father won't worry, and will get well more quickly. Just guess how much I have, Uncle Ben. "

"See here, Ralph," said Uncle Ben, with a twinkle in his eyes, "I promise to give you two cents for each cent you have earned." He did not believe Ralph had earned more than a dollar or two at the most. "Now, how much have yon earned!"

When Ralph told that he had actually earned six dollars and ninety-one cents, and explained just how he had earned it, and had made Uncle Ben count it in the privacy of Ralph's own room, his uncle was truly astonished; but how he did laugh!

"You caught me that time!" he cried, "but I will stand by my promise. A bargain is a bargain. How much do I owe you?"

How much did Uncle Ben owe?

2 x $6.91 = $13.82, *amount Uncle Ben owed Ralph.*

"Not a thing. Uncle Ben," said Ralph; "that was just a joke."

But Uncle Ben insisted, and when he took out a pocketbook almost as fat as the one Ralph had found in the post office Ralph consented to take the money.

When he added it to his twelve dollars and sixty-two cents he was so happy that he fairly danced for joy.

How much had he in all?

$12.62 +$13.82 =$26.44, *total amount in bank.*

"Come in here, please. Mother, "he called. "Here is your surprise 1 I told you it was a good one!"

Mrs. Merton and Aunt Lucy went into the little room and Ralph excitedly told them all about how he had earned the money to help pay the doctor and the grocer.

"It is the biggest and best surprise I have ever had in my life!" said Mrs. Merton, and then of course she kissed him — as mothers have a way of doing. Aunt Lucy just looked at him admiringly and exclaimed:

"Did you ever! Think of a little boy earning so much money!"

"When a chap can pick up money like that he should not be called a little boy: he is a man!" said Uncle Ben. Then turning to Ralph, he said: "Well, Mr. Ralph Merton, isn't this circus-day? I have enough money left for a couple of circus tickets, and I propose that we go out right now and buy them. You have worked hard enough this week to deserve a treat."

So, while his father slept and his mother and Aunt Lucy worked and visited, Ralph went with Uncle Ben to the circus and enjoyed the sight of the zoo animals and all the wonderful feats and performances.

As Uncle Ben, Aunt Lucy, and Mrs. Merton sat talking, late that night, they heard a sleepy voice from Ralph's bedroom call:

"Uncle Ben, I forgot all about selling my papers to-day. If I had sold eighteen I should have had twenty-seven dollars *even,* shouldn't I?"

"You had better go to sleep," replied Uncle Ben. "You aren't reckoning right."

What mistake did Ralph make when he was sleepy?

He said 18 x $0.03 = $0.56, *which was* $0.02 *too much.*

How much would he have had?

18 x $0.03 =$0.54 +$26.44 =$26.98, *amount he would have had if he had sold papers.*

"Sister Jane," said Uncle Ben solemnly, "when that boy tries to earn money after he is asleep it is time something was done. You had better send him out to the farm next week and let him have the vacation he had planned for. I believe I have money enough to run this house until James is around again, and Lucy and I will stay right here with you. What do you say?"

"I say 'yes' with all my heart. Brother Ben," said Mrs. Merton, "for if ever a boy deserved a vacation, I am sure Ralph does!"

And so it turned out that Ralph did go to the farm, after all.

Ralph's Summer on the Farm

Suggestions

Let each pupil keep a "Farm Book" with Ralph. This offers an excellent opportunity for teaching systematic arrangement. Make the booklets of paper large enough to allow for the proper placing of the bill shown on page 33.

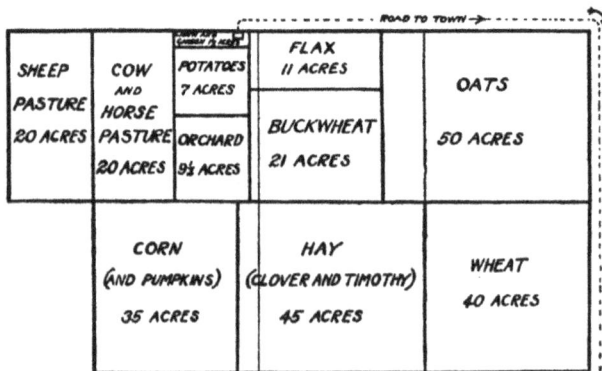

This plan of the farm should be placed on the blackboard for reference during the reading of the story. Make each measurement nine times as long as in the plan dictated on Pages 27-28, i.e., the original oblong 54 in. x 36 in., and the sheep pasture 9 in. x 18 in.

Divide the large oblong into six equal parts, using very light lines. (See above.) Each division represents forty acres. Without attempting absolute accuracy, move the lines a little aside from the original light lines, thus approximating the number of acres in each division.

The children may put this plan into their "Farm Book."

Ralph's Summer on The Farm

I

"How many miles is it to Grandpa's farm, Uncle Ben?" asked Ralph, when he learned the following morning that he was really to go to the farm after all.

Uncle Ben took a time-table from his pocket and after studying it a few minutes drew a diagram on the margin something like this:

Teacher: *Draw on the blackboard, letting one quarter inch represent a mile, to keep the proper relation between the various points:*

HOME	28 mi.	36 mi.	25 mi.	31½ mi.	27 mi.	19 mi.	21 mi.	18½ mi. WHEATLAND

"Each dot," explained he, as he handed it to Ralph, "represents a town. Find out for yourself how far it is."

How did Ralph find out? How far was it?

28 + 36 + 25+31 ½ + 27 + 19 + 21 + 18 ½ = 206, *number of miles from Ralph's home to Wheatland.*

"That's a long way," said Ralph.

The following Tuesday morning Ralph was up and dressed before any one else. By seven o'clock he had finished his breakfast.

"Hurry up, Uncle Ben," exclaimed he, "or we shall be late for the train."

"There's no hurry, young man!" laughed Uncle Ben. "Give me time to finish my breakfast"; and very deliberately he poured the amber syrup over the golden-brown breakfast-cakes and stirred sugar into his steaming coffee. "You had better eat something. You will be hungry before dinner-time."

"I don't feel hungry, Uncle Ben. I suppose it's because I'm so excited over my trip," Ralph replied.

"I will put up a box of lunch for you to eat on the train," said his mother, "for you will not reach Grandpa's until after four o'clock."

"For how many hours shall I ride on the train?" asked Ralph.

"The train leaves here at eight-twelve and arrives at Wheatland at four-twenty-five. How many hours do you make itt" Uncle Ben did not believe in telling children anything they could find out for themselves.

Ralph considered a while and then what do you suppose he said?

Let the children give their answers in their own ways. Who uses Ralph's way?

"If the train got in at four-twelve," Ralph said, "it would take exactly eight hours, but it doesn't get in until twenty-five minutes past four, so that makes thirteen minutes more. In all it would be eight hours and thirteen minutes. That's a long time to ride! I'm glad, though, for I like to ride in the cars find see everything go whizzing past. It's just like a moving-picture show."

At last Uncle Ben finished his breakfast and he and Ralph started for the railroad station. As they entered the station Ralph asked:

"How much will my ticket cost, Uncle Ben?"

"You are a regular question-box, Ralph!" cried his uncle. "It's lucky for me that I know how to make you answer a few of your own questions," he added, teasingly. "How far did you find it was to Wheatland?" he asked.

"Two hundred six miles," promptly answered Ralph.

"Well, it costs two and a half cents a mile. Now, you tell me how much your ticket will cost."

What do you suppose Ralph answered?

Try different ways. Ralph's way was this: If the fare were just 2c a mile it would cost 206 x 2c, *or* 412c. *But it costs* ½ c *more on each mile, so that would he* 206 x ½ = 103c + 412c =515 c= $5.15.

"It will cost five dollars and fifteen cents," replied Ralph.

"Right you are!" said Uncle Ben, as he went to the ticket window and passed a ten-dollar bill and a two-dollar bill through it. When he returned he had a ticket with a return coupon attached.

"This part is to go on, and this to come back on," he said, showing Ralph each part in turn. "Be careful not to lose this coupon or you may have to walk back." He gave the ticket to Ralph and also handed him all the change the agent had given him. "Put this change into your pocket," he said. "You will need some spending-money when you go to town with Grandpa."
How much did Uncle Ben give Ralph?

2 x $5.15 = $10.30, *cost of ticket both ways.*

$10.00+ $2.00 =$12.00, *amount given to agent.*

$12.00 - $10.30 =$1.70, *change received by Ralph.*

"Oh, thank you, Uncle Ben!" cried Ralph. "When I 'm a man I will pay you back, and buy you an automobile besides!" he added. An automobile was the most desirable thing of which his mind could conceive.

Presently the big engine came puffing in, and in a few minutes Uncle Ben had seen Ralph safely seated in one of the cars and had said good-by, and the train was flying along past houses and barns and fields and hills and rivers and ponds, every minute farther and farther from home and nearer and nearer to the farm.

II

Ralph enjoyed his ride exceedingly. However, he was glad when he heard the call:

"Wheatland the next station!" and when a little later the train slowed down and the brakeman opened the door and shouted: "W-h-e-a-t-l-a-n-d!" he gathered up his belongings and hurried to the vestibule, so that he might be sure to get off in time. He stepped down to the platform and there stood

Grandpa, waiting for him. When Ralph shouted, "Grandpa! Grandpa!" and gave him a bear-like hug, people smiled. Probably they were thinking of the time when they weren't grown-ups and had grandpas of their own.

Soon Grandpa and Ralph were seated in Grandpa's light wagon and Daisy and Dandy, his gray ponies, were trotting along at a lively gait.

"How far is it from Wheatland to the farm, Grandpa?" asked Ralph.

"Just eight miles; but it doesn't take long to get there with this team," replied Grandpa. "They are good travelers."

"Let 's see," said Ralph, "if I had to pay you the same rate that the railroad charges, how much money should I have left?"

How much money would Ralph have had left?

8 x $0.025 = $0.20. $1.70 - $0.20 = $1.50, *Ralph would have had left.*

"But this doesn't cost anything and it's even more fun than the cars," Ralph added.

Practically the last mile of the drive was along the edge of Grandpa's farm. The broad acres, thriftily cared for, lay spread out before them, and on a little rise of ground surrounded by wide-spreading oak trees stood the comfortable old farmhouse. At some distance in the rear the big red barn, the granary, the chicken-house, the pig-pen, the corncrib, the silo, and the tall windmill grouped together made Ralph think of a little town.

"It *is* a little town just for animals," he thought.

"How long have you lived on this farm. Grandpa?" he asked, as his eyes roved over the broad acres.

"Let me see," said Grandpa, musingly. "Your grandmother and I took up this homestead when we were first married. It was the first of May, eighteen hundred seventy-seven, when we moved into our little cabin. That was just —

"Don't tell me!" Ralph cried eagerly, "let me find out for myself, please! Uncle Ben always does. This is the twenty-second of June, nineteen hundred fifteen."

How did Ralph find out how many years Grandpa had lived on the farm? Is there more than one way? Can you find Ralph 's way?

Ralph's way: From 1877 *to* 1900 *would be* 23 *years, and* 15 *years more would be* 38 years. *From May* 1st *to June* 22d *is* 1 *mo.,* 22 *da. So it was just* 38 *yr.,* 1 *mo.,* 22 *da. that Grandpa had been there.*

"That's a fact," said Grandpa, when Ralph gave the answer; "but it doesn't seem that long. How time does fly!"

"How old were you when you were married?" asked Ralph.

"Just turned twenty-one. People used to marry younger in those days than they do now," replied his grandfather.

Ralph thought a minute and then said: "I know how old you are, now, Grandpa. You're just —

How old was his grandfather?

21 *yr.* + 38 *yr.* = 59 *yr. old.*

"You are fifty-nine years old," said Ralph. "My, but you don't look as old as that 1 Your hair is only a tiny bit gray and you look so strong!" he added, with an admiring side glance.

"I feel as young as ever, and can do more work than any young man," laughed Grandpa.

"Oh! what is that pretty blue field? It looks as though a piece of the sky had fallen down on your farm!" exclaimed Ralph as they came to a field of flax in full bloom.

"Flax is the prettiest crop there is," replied Grandpa, as he looked with satisfaction at the blue field. "I have eleven acres of it, and I think it will run ten bushels to the acre."

"Then," said Ralph, "you'll have—"

How many bushels of flaxseed would Grandpa have?

11 x 10 *bu.* = 110 *bu. flaxseed.*

"One hundred ten bushels," said Ralph. "How much do you get a bushel?"

"At the very least two dollars and thirty-five cents, probably more," was the reply.

"I can do that in my head, too," said Ralph. Can you, children t How do you think Ralph did it?

Ralph's way: An even 100 *bushels would come to* $235.00. 10 *bushels are worth* $23.50, *and adding them together makes* $258.50.

"You will get two hundred fifty-eight dollars and fifty cents, "said Ralph. "That's right, isn't it, Grandpa? I like to think out ways of doing examples in my head. Oh! there are Grandma and King!"

For just then the ponies turned into the driveway and in a twinkling they had swung up to the broad veranda where Grandma smilingly waited, and King, the big collie, wagged his tail in joy.

III

After the first greetings were over. Grandma announced an early supper. "Because," she said, "this boy must be hungry after traveling all day, and I will warrant he didn't eat much breakfast."

Have you ever eaten supper on a farm? If not, you cannot imagine all the good things Grandma had for supper that evening. There were white, mealy potatoes with their gray jackets still on; fried chicken, and plenty of it; rich brown gravy; flaky white bread, and a big ball of golden butter; real cream for the grown people 's coffee, and milk almost as rich as cream for Ralph to drink; two kinds of pickles, and big red strawberries smothered in cream, and delicious sponge-cake. Everything was so good that Ralph ate enough to make up for the breakfast he hadn't eaten.

"I should like to live here always, Grandma," he said as he left the table; but immediately he added: "Of course, I should want my father and mother to be here, too."

As they were all sitting on the veranda after the chores were done, Ralph said: "Uncle Ben says I'm a question box. Do you care if I ask a lot of questions! You see, I want to find out all I can about the farm, because I want to tell people at home all about it. Besides, I think I shall be a farmer, myself, sometime, and I should like to know something to start on."

"Ask all the questions you want to, and if I get tired of them I will turn you over to Jerry," laughed Grandpa. "You will help me out, eh, Jerry?" he said to the hired man, who sat on the steps.

"Sure thing! I think I can answer as fast as he can ask," responded the obliging Jerry.

"Asking questions is a good way to learn things; if you ask the right sort, "Grandma observed.

"If you're going to make a regular business of these questions, suppose you write down What I tell you, in a book," Grandpa suggested. "Then you won't have to keep asking the same things over."

"That's a good idea," responded Ralph, "and I will call it my 'Farm Book.'"

When they went into the house and lighted the big lamp Grandma hunted through her desk drawers until she found an old account book which Ralph declared just the thing, and he at once began: "How large is your farm?"

Remembering Ralph's wish not to be told his age, Grandpa began, and Ralph wrote the numbers down in his book.

Children to write them as given.

"There are forty acres of wheat, fifty acres of oats, twenty-one acres of buckwheat, eleven acres of flax, thirty-five acres of com, seven acres of potatoes, nine and a half acres of orchard, forty acres of pasture, forty-five acres of hay, and one and a half acres of lawn and garden. How many acres are there in my farm?"

How many acres were there in Grandpa's farm!

$40 + 50 + 21 + 11 + 35 + 7 + 9 ½ + 40 + 45 + 1 ½ = 260$, *number of acres in farm.*

"Two hundred sixty acres," Ralph answered, after a moment's thought. "My! but that's a big farm! I wonder how many miles of fence it takes to go around it."

"Perhaps you can find that out, too," replied his grandfather. "Ask Grandma for a big piece of paper and you can draw a plan of the farm."

The paper was produced, and also a foot rule.

"Now," said Grandpa, "draw an oblong six inches long and four inches wide.

Children draw as dictated; or teacher draw on blackboard.

"Each inch on the paper means one-eighth of a mile. How long is my farm f Can you answer, children?

$6x ⅛ mi. = 6/8 = 1 mi.$ long.

27

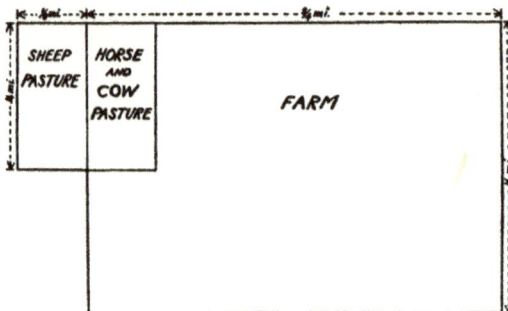

SHEEP PASTURE — HORSE AND COW PASTURE — FARM

"Three-quarters of a mile long," Ralph answered. "How wide is it?" Grandpa asked next; and Ralph again answered correctly. What was his answer!

$4 \times \frac{1}{8}$ *mi.* $= 4/8 = \frac{1}{2}$ *mi. wide.*

"It's half a mile wide," said Ralph.

"Yes, that's right," Grandpa agreed, "except that I bought another little piece for a sheep pasture. To make this extend the top line one inch to the left. Now finish out a small oblong two inches long, using that one inch for the end. Next it make another oblong just the same size, in the corner of the big oblong. That is my pasture for cows and horses. Now, can you find how many miles of fence I have?"

Can *you*, children? How many different ways can you work this out? Which is the shortest way?

Around the big oblong $= 2 \times \frac{3}{4}$ *mi.* $+ 2 \times \frac{1}{2}$ *mi.* $= 2 \frac{1}{2}$ *mi.*

Around the two pastures $= 4 \times \frac{1}{4}$ *mi.* $= 1$ *mi.*

Around the whole farm $= 2 \frac{1}{2}$ *mi.* $+ 1$ *mi.* $= 3 \frac{1}{2}$ *mi.* $- \frac{1}{8}$ *mi.* (*counted twice*) $=$ $3\ 3/8$ *mi. of fence.*

Ralph worked in silence for a few moments; then he cried:

"Why, Grandpa, you have three and three-eighths miles of fence! Didn't your fence cost a great deal of money?"

"Well, yes," his grandfather answered. "It cost about eighty-five dollars for every half-mile, counting posts and wire."

Ralph went to work again and presently he said:

"Then the whole fence cost —

How much? Who can find the very shortest way?

$2 \times \$85.00 = \$170.00$, *cost of 1 mi.*

$3\ 3/8 \times \$170.00 = \$573.75$, *cost of fence.*

"Come, come, that's enough for to-night! You should have been in bed an hour ago," interrupted Grandma. So Ralph followed her upstairs to his room, where she tucked him into his little white bed, and soon he was in the land of dreams.

IV

"Come and see my little chickens," called Grandma from the chicken-house, the next morning. Ralph, who was playing in the yard with King, joined her at once.

"How many are there. Grandma?" he asked.

"They didn't do very well this year," his grandmother replied. "I set five hens and put fifteen eggs under each. The first one hatched ten, the next two twelve apiece, the fourth only eight, and the last one is coming off this morning. We'll see how many she has."

Grandma "shooed" the big white hen off the nest and Ralph counted the downy little yellow balls she had been covering.

"Oh, Grandma! "he cried, "this hen has done the best of all! She has hatched fourteen chicks. Now you have—"

How many chicks had she?

10 + 12 + 12 + 8 + 14 = 56, *number of chicks.*

"But the hens have wasted a good many eggs, I think. They hatched only fifty-six out of seventy-five, so they have wasted — "

How many?

75 - 56= 19, *number of eggs wasted.*

"We never expect them all to hatch," his grandmother explained; "so we don't think of them as really wasted."

"May I take in the eggs while I am here?" asked Ralph. "If I may, I will keep account of them every day in my Farm Book and add them all up before I go home."

"I shall be very glad to have you do it," responded Grandma.

Just then he heard his grandfather calling.

"Come on, Ralph," he said, "If you want to see us shear the sheep."

Away Ralph ran after his grandfather and Jerry. They crossed the big pasture where Ralph counted nine cows, four calves, four horses, and one tiny colt.

"You have a good deal of stock, Grandpa," he observed. "I have counted —

How many head of stock?

9 + 4 + 4 + 1 = 18, *number of head.*

"All I need, and all I want to take care of," replied his grandfather.

Jerry called the colt and its mother so that Ralph could pet them, and the cows and the calves looked at him with such big, kind eyes that Ralph felt they were all his pets.

When they entered the sheep pasture Jerry called, "She-e-e-p! She-e-e-p! She-e-e-p!" and the sheep all came running and crowded about him. He took some feed from his pocket and they ate out of his hand in a most friendly way. Then they were caught, one by one, and held tight while their wool was cut off with big, queer -looking shears.

"Does it hurt them, Jerry?" asked Ralph, as he watched the two men deftly clip the white wool and put it into a big pile which looked like a snowdrift.

"Not a bit," answered Jerry. "They're as glad to get it off as you are to get off your heavy underwear in the spring. They struggle because they don't like being held. See that bare fellow frisking about, over there! Doesn't he look happy? We are late in shearing this year," he added, "because the early summer was rather cool."

"I've counted forty-two sheep, Grandpa. How much wool do you get from each one?"

"These will average about nine and a half pounds a fleece, and this year wool is selling at twenty-six cents a pound," said Grandpa.

"Then you'll make —

How much would he make on the wool?

42 x 9 ½ lb. = 399 *lb. of wool.*

399 x $0.26 = $103.74, *value of wool.*

"Neighbor Graham is anxious to buy three of my sheep, so I suppose I shall have to sell them. I would rather not, but a man must be neighborly," remarked Grandpa.

"How much will he pay you?" asked Ralph.

"Four dollars and seventy-five cents is a fair price, I think," responded his grandfather.

"Then you will get three times four dollars and seventy-five cents, which is —

How much?

3 x $4.75 =$14.25.

"If you add that to your wool," said Ralph, "your sheep will bring you altogether —

How much?

$103.74 +$14.25 =$117.99, *value of wool and sheep sold.*

"That's pretty good," said Grandpa. "In some years the sheep are sick and a good many die. We must take one year with another, in farming."

V

Ralph had told his grandmother how he had earned money to help pay the doctor, so the next morning she asked:

"Don't you want to earn some more money, Ralph, by weeding the garden? I'll pay you five cents an hour if you will do some weeding every day."

"Thank you, Grandma, but I don't need to earn any more money, now that Uncle Ben is helping us," said Ralph. Then he asked: "Will you pay me five apples an hour, instead — the pretty red wealthys, that grow so even and nice!"

"You don't have to work for apples, dear; you may have all you want."

"But, Grandma, I don't want them to eat; and I don't want you to give them to me. I want to earn them."

"Very well, if you like that way better. Then you must have ten apples an hour, for, as we sell them, they bring us only about half a cent apiece," said Grandma, smiling at the eager face of the earnest boy.

"I will keep the account in my book and collect in the fall," said Ralph. "I'm going to try to earn a whole barrelful! How many are there in a barrel?"

"Oh, dear, child, I don 't know!" cried Grandma; "I have never counted them."

"But how many do you think. Grandma?" persisted Ralph:

"Well, the wealthys run pretty even in size and are not very large, so I should guess that there were about five hundred," said Grandma after due consideration. "Yes, I should think there would be five hundred in a barrel."

"Then I'm going to earn five hundred, "said Ralph, with an air of determination.

"Grandma, does Grandpa pay Jerry very much for his work?" asked Ralph later, as he watched Jerry splitting wood and piling it in a long, even pile.

"He pays him twenty-six dollars a month and his board besides," said Grandma. "The board would cost him at least two dollars and a half a week on any farm. He's a good worker and has been with us three years this spring. I really don't know how we should get along without Jerry."

"You've paid him a lot of money," observed Ralph. "Every year you pay him —

How many dollars? How many weeks are there in a year? (Year = 52 *wk.*)

12 x $26.00 =$312.00, *wages for year.*

52 x $2.50 = $130.00, *board for year.*

$312.00 + $130.00 =$442.00, *Jerry earns per year.*

"If he stays through this fall, he will have been here three years and a half. Then in all you will have paid him —

How much?

3 ½ x $442.00 = $1,547.00 *paid Jerry in* 3 ½ *yr.*

"Fifteen hundred dollars!" exclaimed Ralph. "Jerry is almost as rich as Grandpa, isn't he!"

VI

Every morning, after milking the cows, Jerry turned the separator; and almost every day either he or Grandma turned the revolving chum that was so interesting. Ralph frequently helped in this latter operation, and later watched his grandmother mold the big mass of butter into smooth golden balls and then pack them carefully away, wrapped in thin, shiny paper, ready to be sold.

"Will you chum this morning, Ralph, and let Jerry pick the strawberries!" asked his grandmother, early one morning. "Grandpa is going to town, and wants to start before it grows hot."

"Of course I will!" cried Ralph. "I like to churn." And he worked so vigorously that it was not long before his grandmother made the cheering announcement that the butter had "come."

"Get a paper and pencil, Ralph, and make a list of the things I want your grandfather to buy for me in town," said Grandma, as she stood patting the butter into shape. As she directed, Ralph wrote:

Children write list, to the left of the paper; or one pupil write on blackboard.

15 lb. granulated sugar
1 sack Wingold flour
½ gal. molasses
1 lb. Japan tea
2 lb. Mocha coffee
2 gal. vinegar
3 lb. lard
2 sacks table salt
1 side best bacon

"I'm going to send in thirty-five pounds of butter, twelve dozen eggs, and six quarts of strawberries," said Grandma. "Grandpa will take the big wagon, and you may go with him if you like."

If he liked! Ralph was more than delighted at the prospect, and soon he was perched up on the high seat beside Grandpa.

It was an ideal day. The sky was cloudless, and looked like a big blue bowl turned upside down over the earth; the birds sang in the trees and the crickets chirped in the grass. It was sheer happiness to be riding to town behind Daisy and Dandy on so glorious a morning!

"That's the Wilkins farm," explained Grandpa, as they were passing a large, rambling farmhouse. "Now we are exactly six and a half miles from town."

Some time later, they saw a boy walking along the road ahead of them, wheeling a bicycle.

"There's George Wilkins now," said Grandpa, and as they caught up with the boy he called out:

"What's the matter, George? Have you broken down?"

"Yes, sir," replied George. "I punctured my tire when I was a mile and three-quarters from home. I've walked one mile and a quarter since. It's pretty warm walking, too."

"Climb in and ride the rest of the way. There's plenty of room for you and your wheel, too, in this big wagon. That will save you a walk of —

How many miles?

If ¾ mi. + 1 ¼ mi. = 3 mi. George had already come.

6 ½ mi. - 3 mi. = 3 ½ mi. saved.

Arriving at the store, they found that the storekeeper was paying fifteen cents a dozen for eggs, twenty-nine cents a pound for butter, and twelve and a half cents a box for berries. "I'm mighty glad to get these," he said. "It's getting late for strawberries."

How much did they receive for the butter, eggs and berries?

35 x $0.29 =$10.15 received for butter.

12 x $0.15 =$ 1.80 received for eggs.

6 x $0.125=$ 0.75 received for berries.

———

$12.70 received for all.

As the following prices are read, let the children add them to the previous list.

They found that sugar was nine cents a pound; flour one dollar and seventy-five cents a sack; molasses forty-two cents a gallon; lard fourteen cents a pound; tea sixty-five cents a pound; coffee thirty-four cents a pound; vinegar twenty-three cents a gallon; salt nine cents a sack; and the best bacon thirty-eight cents a pound. They chose a side weighing five and a half pounds.

The storekeeper's name was John Vorce, and Grandpa's name was William Grover, although I don't believe Ralph could have told his name without stopping to think, for he so seldom heard it. He always called him "Grandpa"; his mother called him "Father," and his grandmother sometimes said, "William," but oftener "Dear."

"Just make out a bill, will you, Vorce?" requested Grandpa, when he had ordered the groceries. "Let's see, to-day is the second of July, isn't it?"

Children, can you make out a bill?

Question the children about the proper arrangement. Rule lines. Observe form.

			Wheatland, July 2, 1916					
Mr. William Grover								
		Bought of JOHN VORCE						
July 2	15 lb. granulated sugar		@ $	.09	1	35		
" "	1 sack Wingold flour		@	1.75	1	75		
" "	½ gal. molasses		@	.42		21		
" "	1 lb. Japan tea		@	.65		65		
" "	2 lb. Mocha coffee		@	.34		68		
" "	2 gal. vinegar		@	.23		46		
" "	3 lb. lard		@	.14		42		
" "	2 sacks table salt		@	.09		18		
" "	5½ lb. best bacon		@	.38	2	09		
							7	79
	Rec'd Pay't							
	John Vorce							

"Now," said Mr. Vorce, going to the cash register, "I owe you —

How much did Mr. Vorce owe?

$12.70 -$7.79 =$4.91 *amount due Mr. Grover.*

"That's good. Grandpa!" said Ralph. "You've paid for all your groceries and you are taking money home, besides."

With some of his spending-money Ralph bought a flag for the Fourth of July, and then they started for home

VII

As they drove out of town Ralph observed the long line of telephone poles extending along the roadside as far as his eye could reach.

"It takes a whole tree to make a telephone pole, doesn't it?" he asked.

"I suppose so," absently replied Grandpa.

"I'm going to count the poles and see how many trees were used from Wheatland to the farm."

"There's an easier way than counting," suggested Grandpa. "Telephone poles are set ten rods apart. It is eight miles from town to the farm. How many rods are there in a mile?"

"I don't remember," Ralph answered.

Do you remember, children?

"You oughtn't to forget that," said Grandpa. "There are just three hundred twenty rods in a mile. Now can you tell how many poles there are?"

Ralph took his Farm Book from his pocket, opened it on his knee, and although the wagon jolted about considerably, he managed to write —

What did he write?

Let one pupil show at the blackboard:

8 x 320 *rd.* = 2,560 *rd. from Wheatland to farm.*

2,560 *rd.* ÷ 10 rd. = 256. 256 *poles . .256 trees.*

Shorter way: 320 *rd.* ÷ 10 rd.= 32. 32 poles in 1 *mi.* 8 x 32 *poles* = 256 *poles in* 8 *mi.*

Grandpa showed Ralph the shorter way.

"A person would save a lot of time by always doing examples the shortest way," said Ralph. "It makes you think hard to do it, but it pays."

"Aren't there about as many apple trees in your orchard as there are telephone poles from here to the farm?" asked he, a little later.

"As you like so well to find out things, you'd better count them when you get home," replied Grandpa, jokingly.

So as soon as he reached home Ralph went down to the orchard. He saw that the rows of trees were straight; so he counted one row and found it contained twenty-three trees. Then he counted the rows and found there were sixteen. That made —

How many apple trees in the orchard!

16 x 23 trees =368 *trees in orchard.*

"Grandpa, "called he, as he drew near the house, "how many bushels of apples does one tree yield?"

"All trees don't yield alike; but they will easily average nine and a half bushels to a tree in a season like this," his grandfather replied.

"So," said Ralph, "you'll have, in all—"

How many bushels of apples would Grandpa have?

368 x 9 ½ *bu.* = 3,496 *bu. in all.*

"But you sell them by the barrel, don't you? How many bushels are there in a barrel!"

Do *you* know, children!

"Three," answered Grandpa. "I'm surprised that as clever a boy as you shouldn't know that!"

"I forget sometimes," acknowledged Ralph. "Maybe I'll remember better now that I am doing examples about real things on a real farm." In a few minutes he looked up and said: "Then you'll have —

How many barrels of apples?

3,496 *bu.* ÷ 3 *bu.* = 1/165 1/3 *times*. 1,165 1/3 *bbl.*

"Better leave that one-third barrel off, which makes it one thousand one hundred sixty-five. I shall keep three barrels for our own use. How much shall I get for the rest at one dollar and ninety-five cents a barrel? That's about what I expect."

How much would he receive?

1,165 bbl. - 3 bbl. =1,162 *bbl. of apples to sell.*

1,162 x $1.95 = $2,265.90 *will be received for them.*

"That's the largest amount we've reckoned, "
said Ralph.

"Yes, apples will pay well this year," declared Grandpa with satisfaction.

VIII

Some days later a rain set in about noon, so Grandpa and Jerry could not work out-of-doors. Ralph and his grandfather sat on the veranda, looking out across the field of flax, now turning brown, to the big field of oats rapidly changing from green to gold.

"I wonder how much that crop of oats will bring you, Grandpa?" mused Ralph. "It's a good crop, isn't it?"

"Well, yes, it's a pretty good crop, though I have had better. Some years oat crops run thirty bushels to the acre. I calculate this will average about twenty-six."

"And there are fifty acres. That will make —
How many bushels?

50x26 *bu.* = 1,300 *bu. oats.*

"How much does it take for the horses through the winter?"

"I think about five hundred seventy-five bushels will do," answered his grandfather.

"Then you will sell —
How many bushels?

1,300 *bu.* - 575 *bu.* = 725 *bu. oats to sell.*

"Of course, your next question will be how much a bushel, so I will tell you, before you ask, that I expect to get forty cents," laughed Grandpa.

"Then you will get —

725 x $0.40 =$290.00 *expected for oats.*

"Do I ask you too many questions. Grandpa?" inquired Ralph, seriously.

"Not at all! Not at all! I like to see you so much interested in my farm," his grandfather declared, giving him a friendly pat on the head. Ralph was silent for a while; then he said:

"I've been looking at the fence posts and they seem to run together down at the oat field. How far apart are they, really?"

"I always put my posts ten feet apart," replied Grandpa; "then my fence never sags."

Ralph took his Farm Book from his pocket and turned back to where he had reckoned that there were three and three-eighths miles of fence. This time he remembered how many rods there are in a mile.

How many are there?

Knowing how many rods there are in a mile, he could find out how many rods of fence there were around his grandfather's farm.

How did he find this out?

3 3/8 x 320 *rd.* = 1,080 *rd. of fence.*

He had to think hard before he remembered how many feet there are in a rod.

How many are there? (*Rod* = 16 ½ ft.)

He next found how many feet of fence there were, and then how many posts. How did he do this?

16 ½ x 1,080 *ft.* = 17,820 *ft. of fence.*

17,820 ÷ 10 *ft.* = 1,782 *times . 1,782 fence posts.*

"How easy it is to multiply and divide by tens or hundreds or thousands!" said Ralph. "You just don't have to think at all." Then he thoughtfully added: "But I really like the hard ones better. They make me feel as though I had done something — just as I do when our side beats at football."

"That is true," agreed his grandfather. "The harder things are, the more we respect ourselves when we've done them; and every time we do one hard thing we find the next easier."

"I'm glad you said that, Grandpa!" cried Ralph. "I'll try to remember it every time I have a real hard lesson to get. — Oh, look at the rainbow!" he shouted, running down off the veranda; for just then the sun peeped out from behind the clouds, and shining in the misty air produced a wonderful arch in which seven colors were plainly visible.

Name the colors in order, beginning at the top.

"If I could get to the end of it before it faded I should find a pot of gold and a pot of honey."

"Perhaps so," responded Grandpa; "but if you get to the chicken-house before Grandma does you'll be surer of finding something."

"I know what — a basket of eggs!" and Ralph rushed into the kitchen for his basket.

IX

A few rods from the house there was a long row of magnificent cherry trees. Ralph had been watching the gradually ripening fruit with great interest, so he was very glad when he heard his grandmother gay, one morning at breakfast:

"If we don't pick those cherries soon, the robins will. I like the robins, but I object to their eating up my cherries."

"Let me pick them, Grandma," begged Ralph. "I can take the little ladder and climb up into the tree."

36

"Well, I will pay you a cent a quart for all you pick," replied his grand-mother.

"That means two apples a quart," said Ralph. "I'm lucky to have this work; for in August the weeds don't grow very fast, and I haven't earned my five hundred apples yet. This will help me out."

Cherry-picking seemed very delightful work to Ralph, surrounded by the green rustling leaves and the ruby-colored cherries, with the soft breeze blowing and the robins singing their merry songs all about him. How he enjoyed it!

He put the cherries into a basket strapped to his waist, and when the basket was full he clambered down the ladder and filled up the little boxes standing ready on a long table which had been placed beneath the trees. The boxes, when full, were packed into crates.

It was not done in a single day, and, you may be sure, the robins got their share. However, when the last ripe cherry was secured, Ralph found he had picked one hundred sixteen quarts in all.

"I will put up about twenty-eight quarts and send the rest to the store," said Grandma. "I probably shall receive nine cents a quart for them."

"Then," said Ralph, "your cherries will bring —

How much would they bring?

116 *qt.* - 28 *qt.* =88 *qt. to be sold.*

88 x $0.09 = $7.92, *amount cherries will bring.*

"Those plums along the fence by the bam will be ripe before long," remarked Grandma. "Last year we sold twenty-eight crates, besides twelve quarts of the early ones. We sold four quarts for a quarter. I hope we can get as much this year."

Ralph knew that in a crate there were —

How many quarts?

Crate =16 *qt.*

So he soon found how much his grandmother hoped to get for her plums. Can you find out? How?

28 x 16 *qt.* = 448 *qt.* + 12 *qt.* = 460 *qt. in all.*

460 *qt.* ÷ 4 qt. = 115 *times.* 115 *quarters she would receive.*

115 quarters ÷ 4 quarters =28 ¾ times =$28.75, *she would receive.*

"I can do that another way," said Ralph to himself.

Can you? Try.

$0.25 ÷ 4=$0.06 ¼ received for 1 *qt.*

460 x $0.06 ¼ = $28.75 received for all.

He thought a while and then said, "I can do it still another way."

Can you? Try again.

If 4 *qt. sell for* $0.25, 16 *qt.* will sell for 4 x $0.25= $1.00 *received for* 1 *crate.*

28 x $1.00 = $28.00 *received for* 28 *crates.*

12 *qt.* = 12/16 of a crate = ¾ crate.

¾ x $1.00 = $0.75 + $28.00 = $28.75 *received for all the plums.*

"It's queer how many different ways there are of doing some problems, if you just think hard," mused Ralph.

X

"This promises to be an extra good crop of potatoes," said Grandpa to Jerry, as they were spraying the vines with Paris green, while Ralph looked on. "I think they will run one hundred twelve bushels to the acre."

"And there are seven acres," remarked Ralph, "which means that you will have —

How many bushels?

7 x 112 *bu.* = 784 *bu. potatoes.*

"How many bushels do you keep in the cellar?" asked Ralph. "Oh, about fifty-two," answered his grandfather. "Then you will sell —

How many bushels?

784 *bu.* - 52 *bu.* =732 *bu. potatoes to sell.*

"We pay sixty-five cents a bushel for them at home, "said Ralph. "Do you get that much?"

"No, I wish I did; but the storekeeper has to have his profit. I probably shall receive about forty-seven cents."

Ralph took his little Farm Book from his pocket and, sitting down beside a hill of potatoes, he found, after working a minute or two, that his grandfather probably would receive for his potatoes —

How much?

732 x $0.47 =$344.04, *value of potatoes sold.*

"That seems like a lot of money," said he. "But you must remember it costs a good deal to raise them," explained his grandfather. "So it is not all clear gain."

While they were sitting about the table that evening. Grandpa said:

"Get out your Farm Book, Ralph, and we will finish up this estimate of the crops, for I'm certain you will not be happy until it is all set down in your book."

"I'm sure you get just as much enjoyment out of that estimating and figuring as Ralph does," remarked Grandma, who sat knitting warm stockings for Grandpa to wear the coming winter.

"Maybe I do, "acknowledged Grandpa cheerfully.

"Let 's take the wheat crop first, "said Ralph, book and pencil in hand.

"Well, there are forty acres of it," Grandpa began. "It will surely yield thirteen bushels to the acre, and I shall get at least eighty-seven cents a bushel. How much will that be?"

40 x 13 *bu.* = 520 *bu. wheat.*

520 x $0.87 = $452.40, *value of wheat crop.*

"Now the corn," said Ralph.

"There are thirty-five acres of corn," Grandpa responded. "I am sure of twenty-six bushels to the acre. I must keep two hundred forty-five for the stock and I expect to sell the rest at fifty-seven cents a bushel. How much does that figure up?"

35x26 *bu.* = 910 *bu. corn raised.*

910 *bu.* 245 *bu.* = 665 *bu. corn to sell.*

665 x $0.57 = $379.05, *value of corn crop sold.*

"Next, the buckwheat," said Ralph.

"Say nineteen bushels to the acre at seventy cents a bushel. "

"And there are twenty-one acres," said Ralph, looking at his plan. "That will amount to —

How much?

21 x 19 hu. = 399 *bu. buckwheat.*

399 x $0.70 =$279.30, *value of buckwheat.*

"And now the hay. Grandpa."

"That hay," said his grandfather, "will certainly bring me thirteen dollars an acre, and there are forty-five acres. I will sell half and keep the other half for the stock. So I shall make —

How much?

45 x $13.00 = $585.00, *value of entire crop.*

$585.00 ÷ 2 = $292.50, *value of part sold.*

"My book is pretty nearly full," remarked Ralph; "and I know a lot more about farming than I did when I came."

XI

It was Monday, the thirtieth of August, and Ralph was about to take his last ride to town — at least the last one until he was taken to the train. The first day of September he must leave the beloved farm, for school was to begin the following Monday.

"We will drive around by the orchard and cornfield, so that you may take a last look at them, "said his grandfather, as Ralph clambered up into the high seat of the big wagon. So they drove slowly past the long lines of trees laden with apples of varying tones of crimson and gold and green.

"There is no more beautiful crop than the apple crop," observed Grandpa. "We shall begin picking next week. I wish you could be here to help us."

"Don't I, though!" cried Ralph. "But I can't. School begins next week. I'm afraid I shan't get my lessons very well, for I shall be thinking of you and Jerry, out in the orchard, picking apples."

As they drove along the edge of the cornfield, Ralph was astonished at what he saw from his high perch. Scattered all over the field were hundreds of big golden pumpkins.

"There are enough to make a whole army of Jack-o'-lanterns!" he exclaimed. "How many do you suppose there are?"

His grandfather looked critically over the field and then replied:

"There must be almost a thousand — say nine hundred fifty. It's a pretty crop, almost as pretty as the apple crop."

"What do you do with them all?" Ralph asked. "You surely can't make them all into Jack-o'-lanterns and pies!"

"I feed about half of them to the stock in winter. Grandma uses, perhaps, twenty-five of them for pies, and the rest I sell for, say, eight cents apiece. "

"Then you will get for your pumpkins —
How much?
950 ÷ 2= 475 - 25 = 450 *pumpkins to sell.*
450 x $0.08 = $36.00 *to be received for pumpkins.*
"That isn't very much, "remarked Grandpa, "but every little helps."

XII

They drove on to town, where they went first to the store. "Here's a letter for you, Ralph," said the storekeeper, who also was the postmaster. The post office was in one end of the store, as is often the case in very small towns.

Ralph hastily tore open the letter and then fairly shouted:

"Oh, Grandpa, I can help gather the apples! I can help you and Jerry! There is an epidemic of scarlatina in the city and the schools aren't going to open until the twentieth of September; so I can stay two whole weeks longer. Goody! Goody!" and he danced up and down like a jumping-jack.

"Aren't you ashamed of being so glad that there's an epidemic of scarlatina!" exclaimed his grandfather, pretending to be shocked.

"I'm not glad about the epidemic, "cried Ralph; "but I am glad about the apples! Aren't you?"

"Well, yes, my boy; I'm glad you will be with us a while longer. Grandma and I shall be pretty lonely when you're gone," admitted his grandfather.

"I think I'd better reckon up my hours of weeding, to see if I have earned my five hundred apples," said Ralph on the way home. So after supper out came the Farm Book, and he copied all the hours he had worked, by weeks and months. They were:

Teacher: Dictate, or put on blackboard the following:
Hours worked in June: 1 ¾ + 3 ¼ + 2
Hours worked in July: 2 ½ + 5 ½ + 4 + 6 ½ + 5
Hours worked in August: 3 ½ + 5 ½ + 4 + 2 ½
How many hours had he worked, in all?
Adding these by months, the amounts are, — (June) 7 hr. + (July) 23 ½ hr. + (Aug.) 15 ½ hr. = 46 hr. in all.
46 x 10 apples = 460 *apples earned.*

"That isn't enough," declared Ralph. "I need forty more."

"Look up your cherry-picking record," suggested his grandmother. "You earned some apples by picking cherries."

"Oh, yes!" cried Ralph, and he found that he had picked one hundred sixteen quarts of cherries at the rate of two apples a quart. "That, "he said, "amounts to —

40

How many apples?

116 x 2 *apples* = 232 *apples earned.*

"It's more than enough," said he. "Now it's —

How many apples?

460 *apples* +232 *apples* =692 *apples in all.*

"Maybe the extra one hundred ninety-two will pay for a barrel to put them into, "said he delightedly, and his grandmother assured him that there was no doubt about it.

"Now, Grandma," Ralph went on, "I'm going to reckon up the eggs. I began to keep account the twenty-fourth of June and that makes just ten weeks less two days. The hens did much better in July than in August."

"They always do," said Grandma, "because they generally molt in August."

"What's that?" asked Ralph.

"They drop their old feathers and begin to get new ones for the winter," replied Grandma.

"Maybe you would like to keep this account; so I will copy it nicely for you," said Ralph, and this is the way he did it:

Teacher: Copy on blackboard.

First week	304	Sixth week	283
Second week	356	Seventh week	257
Third week	363	Eighth week	224
Fourth week	337	Ninth week	231
Fifth week	295	Tenth week, less 2 days	186
Total, 5 wk	1,655	Total, 5 wk., less 2 da.	1,181

1,655 eggs + 1,181 eggs = 2,836 eggs.

"We reckon eggs by the dozen, Ralph. How many dozen are there in all?" his grandmother asked.

How many dozen were there?

2,836 eggs ÷12 eggs = 236 1/3 times. 236 1/3 *doz.*

"Now, how much are they worth at fifteen cents a dozen?" persisted his grandmother.

236ix $0.15 = $35.45, *total value of eggs.*

"How many hens are there, Grandma?" asked Ralph.

When told there were fifty-eight, he found out how much each hen had earned.

How did he do it?

$35.45 ÷ 58 = 61 7/58 *times* . $0.61: 7/58 *each hen had earned.*

"That isn't much, but thirty cents a month is pretty good for just a hen to earn," said he.

XIII

"I really must find out how much money Grandpa expects to get out of all his crops," said Ralph, "and then he must write and tell me if it comes out that way. And that will probably be the very last thing in my Farm Book. "

So he set the values of the crops all down in a column like this:

Teacher: Dictate or place on blackboard:

Value of flax..$ 258.50

Value of wool..103.74

Value of apples..2,265.90

Value of oats..290.00

Value of potatoes..344.04

Value of wheat..452.40

Value of corn..379.06

Value of buckwheat..279.30

Value of hay..292.50

Value of pumpkins..36.00

Total..$4,701.43

"I didn't add in the cherries, plums, and eggs, because they belong to you. Grandma," he explained. He looked at the figures critically, and then closing his book, remarked: "Farming is the best business there is. I'm almost sure I shall be a farmer when I grow up."

How delightful the summer had been! Everyday there had been something different for Ralph to do. Sometimes he had helped harness the horses, and often he had ridden Dandy bareback to the pasture. Many a game of hide-and-seek he had had with King in the orchard. On rainy days, out in the big, clean barn, Jerry had told him most thrilling stories of adventure, for Jerry was "just full of stories," as Ralph expressed it. When they cut the grain Ralph had ridden on the self-binder with his grandfather and it was fascinating to watch the bundles of grain drop so regularly from the big machine. He had also ridden on the hay-rake, which was quite as interesting. He had fed the pigs and the chickens and had gathered all kinds of fruit. In short, he had done all the delightful things that a boy can do nowhere except on a farm. And he had actually gained twelve pounds!

And now the apples were gathered, the com was shocked, the grain was harvested, and the last day of Ralph's stay on the farm had come.

"I've been here a long time," said he as he sat on the veranda steps, looking thoughtfully out over the harvested fields. "I'm going to find out just how many times I have slept here." He remembered that he had arrived at the farm on the twenty-third of June, and today was the fourteenth of September.

How did he find out how many nights he had slept at the farm?

From June 23rd to June 30th = 8 nights.

From July 1st to July 31st = 31 nights.

From August 1st to August 31st = 31 nights.

From September 1st to September 14th = 14 nights. 8 + 31 + 31 + 14 = 84 nights in all.

"I've eaten three meals every day, besides the cookies and doughnuts that Grandma has given me between meals. That makes —

How many meals?

84 x 3 *meals* = 252 *meals in all.*

"No, that's not right," Ralph went on. "The first day I ate only my supper here. That makes the number of meals two hundred fifty. If I had had to pay what would be charged for such good meals in the city, they would have cost — how much do you suppose. Grandma?"

"At least twenty-five cents a meal," said Grandma.

"Then that would have been —

How much?

250 x $0.25 = $62.50, cost of all meals.

"Well! I never should have thought it cost so much just for a boy to eat! I really ought not to take those apples home. Grandma. But I *did* want to surprise Father and Mother with them," added he.

"Don't you worry about those apples!" exclaimed Grandma. "You worked hard for them; and Grandpa shall ship them so they can go right along on the train with you."

So when, the next morning, Ralph climbed up to the high seat beside his grandfather, there, in the back of the wagon, sat the barrel of apples that he himself had earned and picked and helped pack. He looked at it with pride.

"It's much better to work for apples than for money," he declared. "Apples look so much bigger! Next summer I shall be a whole year older and maybe I can earn two barrels. Good-by again. Grandma!" he called, as the horses started down the driveway. "I will surely come back next summer."

And away Ralph went, leaving his grandmother wiping her eyes on the corner of her apron, and King looking wistfully after his departing playmate. Ralph's delightful vacation was over!

How Everybody Helped

SUGGESTIONS

Use the title of this story as the name of a booklet to be made by the pupils. Let each problem regarding contributions be written under the specific name of the pupil contributing, as well as the number. A name means much more to a child than a number.

The different methods of arriving at the amounts donated should be freely discussed. Commend short and original methods.

HOW EVERYBODY HELPED

It was a beautiful day, that twentieth of September on which school reopened. The scarlatina patients had all recovered and the children, especially those who had been quarantined, were happy at the thought of going to school after their vacation. For once, the holidays had seemed too long. Most of the children had been promoted, and everybody knows how interesting it is to move on to a higher room, to meet an entirely new teacher, and make a fresh start.

When Ralph returned home from his grandfather's farm he found his father almost as well as ever. Indeed, the doctor had said Mr. Merton might resume his work the following week; and as Ralph had felt no responsibility for the support of the family since Uncle Ben had taken charge, the boy was very light-hearted. His schoolmates were all glad to see him again, and he had much to tell them about his good times on the farm.

"There are a new boy and girl for our class," said Joe Beverley, as he and Ralph were about to start from Ralph's home for school on Monday morning. "They're twins and they are from Chicago. They live right next door to us. I like them very much. There they come now!" exclaimed he, as a boy and a girl came around the comer. "Come on over here, Floyd," he called. "I want you and Flora to meet my chum; and Ralph was soon becoming acquainted with the Aldrich twins.

A few days later, on the way home from school. Flora Aldrich said to Ralph:

"You don't seem to have much in your schools here. In our school in Chicago we had a victrola and a radiopticon and other interesting things. Why don't you get something like that, to make school pleasanter?"

"What's a radiopticon?" asked Ralph.

"It's a machine that makes big pictures on a screen," explained Flora. "You can put postcards and other pictures into it and the same pictures will be shown much bigger on the screen, in a room that is all dark. We used to take such nice journeys with it. We 'd start with a picture of a train or a steamer and pretend we were going to a certain country. We 'd have all the pictures of that country shown in their regular order. My, but those trips made us all like geography! And they made us understand it, too."

"Yes, and we had Wild West pictures of cowboys and Indians, and all kinds of animals.; and sometimes pictures from 'The Wonder-Book' and the King Arthur stories; and funny ones from 'Alice in Wonderland,'" added Floyd.

"That must have been fun," replied Ralph. "I suppose we shall have one when we get around to it." Then he loyally added: "Our school can get anything that any other school can get. I will ask Miss Overton about it tomorrow."

The next day Ralph went to school earlier than usual Before the other children arrived he had time to tell his teacher, Miss Overton, all about the radiopticon and the great interest it would bring to the school work, according to the Aldrich twins.

"Mayn't we have one. Miss Overton?" he asked.

"I wish we might, Ralph," she replied, "but where could we get the money for it? A good radiopticon would cost considerable money and of course we must have a good one if we have one at all."

Ralph's eyes brightened.

"Why can't we boys and girls each earn some money and put all that we earn together and buy one?" he said. "I know a good many ways of earning money"; and he told her of what he had done to help when his father was hurt.

"Your idea is a fine one, Ralph!" said Miss Overton. "Suppose you go and talk it over with Miss Fowler. If each boy and girl in the building would help even a little, we could surely get a radiopticon."

Ralph went to the office of the principal, Miss Fowler, and his earnest, manly way of presenting his idea at once won her sympathy.

"I like your suggestion of the children earning the money, and not just asking their parents to give it to them," she said. "Then, when it was all earned, each pupil could tell his own room just how he had earned his contribution."

"I think we should enjoy that, Miss Fowler," eagerly responded Ralph. "We could take a week or two to earn the money, and have an afternoon in which to tell how we did it."

"Yes, we could call it an 'Experience Meeting,'" smilingly suggested Miss Fowler.

The children were not only willing but eager to help, and they all thought, with Ralph, that a week or two would be time enough in which to earn the necessary money. However, the teachers convinced them that it would be much better to take a longer time. So they finally decided on the second week in November as that in which to report.

In the Sixth Grade the children organized into two bands, the Hustlers and the Rustlers. Ralph was captain of the Hustlers, while Flora Aldrich was chosen head of the Rustlers. The two captains kept quizzing their respective companies as to their progress, and suggesting ways and means, so that no one was allowed to forget for a moment his responsibility to his side. As we have seen, Ralph was a real "hustler," and Flora, with her superior knowledge of how people did things in Chicago, was an excellent "rustler." So excitement ran high in the Sixth Grade.

The children even went so far as to cut numbers from calendars and wear them as badges, to show on which side they were, the Hustlers taking the odd numbers, which they tied with blue ribbon, and the Rustlers the even numbers garnished with red. Blue and red were the school colors.

"Let's ask Miss Overton to call us by number instead of by name, when we give our experiences," Ralph suggested to his side. "And let 's tell what we earned in a way that will make the Rustlers work it out before they can tell how much we have!"

So they worked the matter up among the Hustlers, intending to keep it a "dead secret." But it isn't easy for fifteen lively boys and girls to keep a secret.

Floyd Aldrich, who was on Ralph's side, made a remark which set Flora to guessing, and before he realized it Floyd had given away the "dead secret" to the enemy.

"You Hustlers think you're odd just because you wear odd numbers, but we shall be even with you," said she. "You shall find we don't wear even numbers for nothing." So the Rustlers planned to do the very thing the Hustlers had in mind; and they did manage to keep it a secret.

II

Six weeks seemed a long time, but they finally passed and the great day of the report arrived. An air of intense excitement prevailed throughout the building; for each room was hoping to have the largest amount when the returns were all in.

As soon as school opened on this momentous afternoon, Miss Overton placed on her desk two glass bowls, the kind used for goldfish, one decorated with blue and the other with red ribbon.

"As each one tells how he earned his money, he may deposit it in the bowl bearing his color," said she. "I am using glass bowls so that you can watch your fund grow."

The captains took their places at the blackboard and wrote in a column the odd and even numbers, respectively.

Teacher: Choose two pupils to represent Ralph and Flora at the board, and keep tally.

Divide the school into Hustlers and Rustlers and let each group work their opponents' problems.

"The Rustlers will need pencil and paper. Miss Overton," explained Ralph, as he took his place at the board.

"So will the Hustlers, if they want to find out how much we are giving," cried Flora Aldrich. "They thought they would be *odd* and do something different, but we're *even*," she added, amid much laughter and a scramble for pencils.

"Number One," called Miss Overton, and Floyd Aldrich arose.

"I peddled bills every day for a week," said he. "Monday I peddled one hundred; Tuesday two hundred; Wednesday four-fifths of a hundred; Thursday one and one-fifth hundreds; Friday two and two-fifths hundreds; and Saturday three and three* fifths hundreds. I got fifteen cents a hundred. Out of what I got I used sixty-five cents for a history, bought two pencils at five cents apiece, and an eraser for two cents. The rest I give to the fund"; and he walked up and deposited a handful of small coin in the bowl tied with blue ribbon.

"I suppose the Rustlers must tell how much Floyd has contributed," said Miss Overton.

Children, how much did Floyd contribute?

100 + 200 + 80 + 120 + 240 + 360 = 1100, *number of bills peddled.*

11 x $0.15 = $1.65 *received for peddling bills.*
2 x $0.05 = $0.10 + $0.02 = $0.12 + $0.65 = $0.77 *spent.*
$1.65 -$0.77 =$0.88 *contributed.*

Joe Beverley gave the right answer, and Ralph wrote the amount after Number One.

Number Two was Dennis Ryan, the first on the list of Rustlers, who said:

"I helped Father make a garden last summer and when I told him I wanted to earn some money for school he said I might have some vegetables to sell. So I sold five pumpkins at nine cents apiece [Here the Hustlers began to use their pencils, so that Dennis might not have to repeat as Floyd had done]; eight cabbages at four cents a head, one peck of onions at eighty-six cents a bushel, and a half -bushel of rutabagas at fifty-five cents a bushel. I bought a pair of mittens, for which I paid fifty cents. The rest I deposit for the school."

He suited the action to the word. The amount was correctly given by a Hustler, and Flora made her first record.

How much did Dennis contribute?

5 x $0.09 = $0.45 *received for pumpkins.*
8 x $0.04 =$0.32 *received for cabbages.*
.86 ÷ 4 = $0.21 ½ *received for onions.*
½ x $0.55 = $0.27 ½ *received for rutabagas.*
$1.26 *received in all.*
$1.26 - $0.50 = $0.76 *for the fund.*

"Number Three," said Miss Overton.

"When I wanted to earn some money, I could not think of a way," said little Jean Bryce, "so Mamma suggested that I have an ice-cream sale. She bought two bricks of ice-cream which cost twenty-five cents each, and made a cake which cost her thirty-five cents; not counting her work. Then I told eight of my girl friends about it and they all came; besides there were Papa and Mamma, Grandpa and Grandma, Uncle Rex and Aunt Mary, and my big brother, Billy. I sold ice-cream and cake for ten cents to each of them; and, to be perfectly fair, I took ten cents out of my bank to pay for what I ate myself. Then I paid Mamma the cost of the cream and cake and now give the rest."

The Rustlers soon knew how much Jean had given.

Do you know!

2 x $0.25 =$0.50, *cost of ice-cream.*
$0.50 + $0.35 = $0.85, *cost of ice-cream and cake.*
16 x $0.10 =$1.60 *received for cake and cream.*
$1.60 -$0.85 =$0.75, *amount given school.*

"Number Four," called Miss Overton.

"I baked some cookies for the County Fair," said Genevieve Hart, "and won the first premium, which was fifty cents. Papa said that he 'd give me thirty-five cents if I would bake him a batch just as good; but I paid ten cents back to Mamma for the material I used. Mamma said she thought the money really belonged to the school, anyway, because I had learned in the cooking-class to make cookies."

How much did Genevieve give?

$0.50 + $0.35 = $0.85 - $0.10 =$0.75, *amount given.*

The Hustlers gave this amount without even looking at their pencils, and it was set down.

"Number Five," said Miss Overton.

Walter Brown reported that he had raked leaves and packed them around shrubbery, for three different people, at twelve cents an hour. He had worked one hour and a half one day; two hours another, and one and three-quarters another. Then his father had paid him a quarter for getting a standing above eighty-five in every subject on his report card the month before.

"That was harder work than raking leaves, "said he, as he dropped his earnings into the bowl decorated with the blue bow.

This problem wasn't quite so easy as the last, but in a few minutes a half-dozen Rustlers had found the correct amount, and Ralph set it down.

What was it?

1 ½ hr. + 2 hr. + 1 ¾ hr.-5 ¼ hr. *of work.*

5 ¼ $0.12 = $0.63 *earned by raking leaves.*

$0.63 +$0.25 =$0.88, *amount donated.*

Number Six was next called.

"I bought a shoe-cleaning kit for a quarter," said Carl Cummings, "and shined my big brother's shoes, at a nickel a time, every morning for three weeks, except Sundays. I took out the money to pay for the kit, and now give the rest to the Rustlers"; and he dropped it into their bowl.

The Hustlers announced Carl's contribution, which was —

How much?

3 x 6 *da.* = 18 *da., number of times he shined shoes.*

18 x $0.05 = $0.90 *received for shining shoes.*

$0.90 - $0.25 = $0.65 *donated to the Rustlers.*

"Number Seven," said Miss Overton.

"I took care of a neighbor's baby every day after school, when its mother was sick," said Erna Lawton. "She paid me fifteen cents each day after school, and forty-five cents for Saturday. I went to the moving-picture show twice, at ten cents a time, and bought fifteen cents' worth of candy. The rest of the money I have brought."

The opposing side found that she had brought —

How much?

5 x $0.15 = $0.75 *earned after school.*

$0.75 + $0.45 = $1.20 *earned in all.*

2 x $0.10 = $0.20 + $0.15 = $0.35 *spent.*

$1.20 - $0.35 = $0.85 *donated.*

Number Eight was Sigmund Wolski, who had piled wood for a neighbor for two Saturday mornings for twenty-five cents a morning and had carried in wood every day for a week at five cents a day, including Sunday. He had given his little brother a dime and he deposited the remainder in the bowl decorated with red ribbon.

How much did he put in?

2 x $0.25 = $0.50 *earned by piling wood.*
7 x $0.05 = $0.35 *earned by carrying wood.*
$0.50 + $0.35 = *$0.85 earned in all.*
$0.85 -$0.10 = $0.75 *donated.*

This was quickly announced, and Number Nine was called.

"This report takes in David Winchester, too," said Chris Altman, "but he asked me to act as spokesman.

"David and I went hazel-nutting and we picked seven bushels of nuts. When they were dried and shelled each bushel produced only one and one-fourth quarts. We sold them at eighteen cents a quart and divided the money equally."

At first this seemed a poser, but soon light dawned on several faces, and before long the Rustlers had it worked out.

How much did each boy give?

7 x 1 ¼ qt. = 8 ¾ qt. *of shelled hazel nuts.*
8 ¾ x $0.18 = $1.57 ½ *received for hazel nuts.*
$1.57 ½ ÷ 2 = $0.78 ¾, *each boy's share.*

"But we didn't get the half -cent, so you have to drop that three-fourths of a cent," explained Chris, as Ralph began to write it down just as it was given. So he erased the fraction.

"Number Ten is next," said Miss Overton.

"I ironed dish towels every Saturday for five weeks for half a cent a towel, "said Eva Moreland, "and swept and dusted the living-room four times at ten cents a time. The first week I ironed ten towels, the second eleven, the third nine, the fourth twelve, and the fifth (the week we had company) fourteen. It doesn't amount to very much, but I worked a long time to get it." She put a handful of pennies and five-cent pieces into the Rustlers' bowl, while the Hustlers were finding out how much she had brought.

How much was it?

10 + 11 + 9 + 12 + 14 = 56, *number of towels ironed.*
56 x $0.005 = $0.28 *earned by ironing.*
4 x $0.10 = $0.40 *earned by sweeping and dusting.*
$0.28 + $0.40 =$0.68, *total amount earned and given.*

"Now, Number Twelve," said Miss Overton. "David was Eleven, you know."

"I carried some potatoes and pumpkins down into the cellar for my aunt," Joe Beverley reported, "and she gave me a shilling for every four bushels of potatoes and a quarter of a cent for every pumpkin. There were sixteen bushels of potatoes and twelve pumpkins. She paid me fifteen cents for beat-ing a carpet for her. Then I spent ten cents for carfare and three cents for a pencil. "Joe dropped his contribution into the glass bowl, through the sides of which could be seen a fast-increasing quantity of small coin.

How much had he added?

16 *bu. ÷ 4 = 4 of times he earned* $0.125.

4 x $0.125 = $0.50 e*arned by carrying potatoes.*

12x ¼ c = 12/4 = $0.03 *earned by carrying pumpkins.*

$0.50 +$0.03 + $0.15 =$0.68 *earned in all.*

$0.68 - $0.13 = $0.55 *left for the school fund.*

This took the Hustlers a little longer than the other problems, but they gave the amount correctly, and Number Thirteen was called.

"I sold flowers from my own garden almost every day last summer," said Margaret Whitten, "and made a good deal of money. I had several regular customers. I sold twenty-four bunches of pansies at ten cents a bunch, twenty-seven bouquets of sweet peas at fifteen cents apiece, and nineteen bouquets of nasturtiums at five cents apiece. I put it all into my bank and this fall I spent five dollars for a new dress. I am giving you one-third of all I have left of my flower money."

How much did Margaret give?

24 x $0.10 = $2.40 *received for pansies.*

27 x $0.15 = $4.05 *received for sweet peas.*

19 x $0.05 = $0.95 *received for nasturtiums.*

$7.40 *received for all flowers.*

$7.40 - $5.00 = $2.40 *left after buying dress.*

$2.40 ÷ 3 = $0.80, *amount donated.*

"That's a good deal for a girl to earn," remarked Carl Cummings, as he wrote out the amount contributed.

Number Fourteen was Elate Caswell, who had dressed a doll for a neighbor's little girl, for which the mother had paid her forty-five cents.

"I learned at school how to sew," said she, "so I thought it was a good way to pay back to the school. "

She also took care of the same little girl three times while the mother attended a party, a play, and a concert, and for each time she received fifteen cents.

"Out of the money," she said, "I bought two roses for my mother's birthday at fifteen cents each, and a pretty birthday card for a nickel If it had not been for Mother's birthday I could have given more."

Kate gave — how much?

3 x $0.15 =$0.45 +$0.45 =$0.90, *amount earned.*

2 x $0.15 =$0.30 +$0.05 =$0.35, *amount spent.*

$0.90 -$0.35 =$0.55, *amount given.*

It took but a moment to find the amount Kate had given, and then Number Fifteen was called.

"I carried suit-cases to the station for four different traveling-men," said Ernest Belden. "Two of them gave me a dime each, one gave me fifteen cents, and the other one had two suit-cases, so he paid me a quarter. One day I caught a horse that was just about to run away and the owner gave me two shillings. I paid a dime for carfare, two dimes for writing-tablets, and one cent for chewing-gum. The rest belongs to the Hustlers."

The Rustlers quickly found the amount and it was recorded by Ralph. What was it?

2 x $0.10 = $0.20 + $0.15 = $0.35 + $0.25 = $0.60 *earned by carrying.*

$0.60 +$0.25 =$0.85 *earned in all.*

$0.10 +$0.20 +$0.01 =$0.31 *spent.*

$0.85 -$0.31 =$0.54 *left to give.*

"Number Sixteen," called Miss Overton, and Leonard Munroe arose.

"I picked up old shingles and carried them into a shed, for thirteen cents an hour. I worked two afternoons from four o'clock to six. Afterward I cleaned up the yard for a quarter, and found a penny while I was doing it."

The money was placed in the bowl and recorded for the Rustlers. How much did Leonard give?

From 4 o' clock to 6 = 2 *hr.* 2x2 *hr.* = 4 *hr.* labor.

4 x $0.13 = $0.52 *received for piling shingles.*

$0.52 + $0.25 + $0.01 = $0.78, *amount of donation.*

Number Seventeen was called.

"I made a work-bag of taffeta silk and sold it to Auntie for a dollar, * ' said Sallie MacKay. "I used a quarter of a yard of silk at eighty-five cents a yard, two and a half yards of ribbon at five cents a yard, and one-fifth ounce of violet sachet powder at sixty cents an ounce. I took out the cost of the material and this is the result of my work."

What was Sallie 's contribution?

¼ x $0.85 = $0.21 ¼, *cost of silk.*

2 ½ x $0.05 = $0.12 ½, *cost of ribbon.*

⅕ x $0.60 = $0.12, *cost of sachet powder.*

$0.21 ¼ + $0.12 ½ + $0.12 = $0.45 ¾, *cost of material.*

$1.00 - $0.45 ¾ =$0.54 ¼, *amount received for work.*

"I counted three-quarters of a cent as one cent," explained Sallie, as she saw that the Rustlers were using the fraction. "They say they do that in business, if it's over half a cent."

Then it read, $1.00 - $0.46 = $0.54.

Number Eighteen was Hans Schmidt.

"I found some old rubbers, a pair of old rubber boots, and a worn-out bicycle tire in the shed," he said. "I collected some more rubber from our neighbors and in all got fifteen pounds. I sold it to the junkman for seven cents a pound. I've spent two-fifths of the money, but have brought the rest."

15 x $0.07 = $1.05 *received for old rubber.*

5/5 – 2/5 = 3/5 *of the money left.*

$1.05 x 3/5 = $0.63 *donated.*

Hans's donation was promptly reckoned and recorded. How much did he give?

Peter Lewis was Number Nineteen.

"My uncle has a hickory-nut grove a few miles out in the country," said he, "and I went out one Saturday to pick hickory nuts for him on shares. He gave me one-fourth of all I gathered, which I think was very liberal. I gathered eleven bushels in all, and put aside a bushel and a half of my share for our own use, without shelling them. When the rest were shelled they measured just half as much as before. I sold the shelled ones at forty-two cents a peck. It cost me twenty cents to go to the farm, so I paid that, but the rest goes into the bowl with the blue ribbon on it."

The Rustlers were rather puzzled at first by the complications in this hickory-nut problem, but Miss Overton asked them a question or two and they at length saw the way to work it out.

Can you?

$\frac{1}{4}$ x 11 *bu.* =2 $\frac{3}{4}$ *bu., Peter's share.*

2 3/4 *bu.* – 1 $\frac{1}{2}$ *bu.* = 1 $\frac{1}{4}$ *bu. left to sell.*

1 $\frac{1}{4}$ ÷ 2 = 5/8 *bu., amount left after shelling.*

1 *bu.* = 4 *pk.* 5/8 *bu.* = 5/8 x 4 *pk.* = 2 $\frac{1}{2}$ *pk. to sell.*

2 $\frac{1}{2}$ x \$0.42 = \$1.05 *received for hickory nuts.*

\$1.05 - \$0.20 = \$0.85, *amount deposited.*

"Number Twenty," called Miss Overton.

"All our numbers from twenty to thirty belong together," explained Flora Aldrich, "and we will give that last."

So the rest of the Hustlers were called.

Number Twenty-one was Jimmie O'Hagan, and as he rose he said:

"You won't need any pencils for mine: it's too easy. I carried my brother's lunch to him one day and he gave me a dime. Another day a collie dog came to our house and I knew by the way he whined that he was lost. I looked in the morning's paper, and sure enough I found out where to take him. I believe I walked two miles with that dog. His master paid me fifty cents for taking him home. Here's my cash"; and it fell with a clink into the rapidly filling bowl.

"Easy enough," said one of the Rustlers.

\$0.10 +\$0.50 =\$0.60, *Jimmie's donation.*

Gertrude Blake was Number Twenty-three. She had crocheted two pairs of bedroom slippers and sold them at sixty-five cents a pair. She offered to give the Hustlers fifty per cent of the money received for them.

"What does fifty per cent mean?" asked one boy; but most of the children knew.

Do you know?

"That's easier than Jimmie's, "said Sigmund Wolski. "Of course half of twice sixty-five is just sixty-five cents."

But some of the children had worked it all out, without seeing the short way. What other way of working it out is there?

2 x $0.65 = $1.30 *received for slippers.*

50% = ½

½ x $1.30 =$0.65, *amount given.*

"Wasn't that a waste of time?" asked Miss Overton, and they all agreed that it was.

Number Twenty-five was called.

"I couldn't find anything new to do this fall to earn money, "said Jerome Sanders, "but last summer I went fishing several times. In all I caught two bass that averaged a pound and a half each, one pickerel that weighed a pound and three-quarters, and three pike that averaged a pound and a quarter each. I sold the bass for thirteen cents a pound, and the others for twelve cents a pound. I paid eight nickels for my fishing-tackle and lent the rest to my brother. He paid me back a few days ago; so that's what I am giving to the cause."

"That's a hard one!" exclaimed the impetuous Dennis Ryan; but he settled down to work it out.

Several times the Rustlers had to question Jerome still further about the conditions which he had stated; but finally Dennis shouted: "I know how much he's giving! "and he really did.

Do you?

2 x 1 ½ = 3 *lb. of bass.*

3 x 1 ¼ = 3 ¾ *lb. of pike.*

3 ¾ lb. + 1 ¾ *lb.* = 5 ¼ *lb. of pike and pickerel.*

3 x $0.13 = $0.39, *amount received for bass.*

5 ½ $0.12 = $0.66, *amount received for other fish.*

$0.39 + $0.66 = $1.05, *amount received for all fish.*

8 *nickels* = $0.40.

$1.05 - $0.40 = $0.65, *amount given.*

Number Twenty-seven's donation was easier to find.

"I made a brown linen doily with a wide crocheted lace border," explained Louise Farrel. "It was a lot of work and my mother paid me a dollar and a quarter for it. I have kept thirty-five cents to pay for the material and thirty-five cents to buy our baby a present for his second birthday. I'm giving the remainder."

What was she giving?

$0.35 + $0.35 = $0.70 *kept,* or 2 x $0.35 =$0.70 *kept.*

$1.25 - $0.70 = $0.55, *amount given.*

Number Twenty-nine was Ralph Merton, the captain of the Hustlers.

"Every day for two weeks I worked at Granger's fruit store, sorting fruit, for an hour and a half after school. Mr. Granger paid me twelve cents an hour and sold me three dozen over-ripe bananas for six cents a dozen, which I sold for nine cents a dozen. I bought a cap for sixty-five cents, a spelling-book for twenty cents and a tablet and an eraser for nine cents. All the rest is for the radiopticon."

This was a rather long, although not difficult, problem, and when the Rustlers found what Ralph had made, they looked rather crestfallen, for it was so much more than any amount on their list.

What was it?

1 ½ x $0.12 = $0.18, *amount earned per day.*

2 x 5 school days = 10 *days he worked.*

10 x $0.18 = $1.80 *earned by sorting fruit.*

$0.09 - $0.06 =$0.03, *gain on 1 doz. bananas.*

3 x $0.03 = $0.09, *gain on all the bananas.*

$1.80 + $0.09 =$1.89, *money received.*

$0.65 + $0.20 + $0.09 =$0.94, *amount spent.*

$1.89 - $0.94 =$0.95, *amount left for radiopticon.*

When Twenty to Thirty were called, Dorothy Perkins responded:

"Six of us girls had a little bazaar at my home last evening. We had been making and collecting things for it ever since we planned this meeting. We each invited six people, so there were over forty present. We served tea and wafers to thirty-five people at ten cents apiece."

The Hustlers took their pencils and began to work. Referring to a written list, Dorothy continued:

"We sold seven dressed dolls at twenty-five cents apiece, four candleshades, which we had made at school, for eight cents apiece, twelve crepe paper chrysanthemums at eight cents, and eight roses at seven cents. (We also made the flowers ourselves.) Then we sold three and one-half pounds of fudge at thirty-six cents a pound and several small articles which amounted to eighty-two cents." The Hustlers had to "hustle" in earnest to keep up with her. "The material for everything cost us just four dollars and thirty-seven cents, and the net profit is equally divided among six of us."

It took some time to work Dorothy's problem out, but at last it was accomplished.

Can you give the result?

7 x $0.25 = $1.75 *received for dolls.*

4 x $0.08 = $0.32 *received for candle shades.*

12 x $0.08 = $0.96 *received for chrysanthemums.*

8 x $0.07 = $0.56 *received for roses.*

3 ½ $0.36 = $1.26 *received for fudge.*

35 x $0.10 = $3.50 *received for tea and wafers.*

$0.82 *received for small articles.*

$9.17 = *total receipts.*

$9.17 - $4.37 = $4.80 *net profit.*

$4.80 ÷ 6 =$0.80 *donated by each.*

The children could hardly wait to add up the columns and find out which side was ahead. They hurried so that they made mistakes and had to try it over and over. Finally Miss Overton let them add aloud, in concert, to be sure. *Teacher: Let your class do the same.*

HUSTLERS		RUSTLERS	
1	$ .88	2	$ .76
3	.75	4	.75
5	.88	6	.65
7	.85	8	.75
9	.78	10	.68
11	.78	12	.55
13	.80	14	.55
15	.54	16	.78
17	.54	18	.63
19	.85	20	.80
21	.60	22	.80
23	.65	24	.80
25	.65	26	.80
27	.55	28	.80
29	.95	30	.80
	$11.05 *Total*		$10.90 *Total*

"Oh! we're fifteen cents ahead!" shouted the Hustlers in high glee, and they applauded wildly. Miss Overton always permitted her pupils to applaud when the occasion warranted.

"It's a good way to let off steam," she said to teachers who wondered why she did.

"That's only one cent apiece more for our side," cried Flora, "and we will bring that to-morrow; and then we shall be even! We shall have to keep even because we wear the even numbers."

All the Rustlers agreed to do as their captain had suggested. Then the Hustlers cheered again, because the Rustlers were "game" and because there would be fifteen cents more in the entire fund, which would make in all —

How much?

2 x $11.05 = $22.10, *donation of Sixth Grade.*

"Both sides have done wonderfully well!" declared Miss Overton. "I 'm proud of every one of you! You have shown a good deal of school loyalty and civic pride in this work. I'm sure that when you grow up you will all be good citizens."

"Ralph brought more than anybody else," remarked Jimmie O'Hagan.

"You see, I had so much experience in earning money when my father was ill that it helped me this time," Ralph explained. "I don't believe the Eighth Grade has done any better than we; do you. Miss Overton?"

"I really do not," replied the teacher.

"That money ought to be counted, so we may be sure there's no mistake," suggested Hans Schmidt.

"That is a business-like suggestion," said Miss Overton. "Let each captain choose one from his own side to count his opponent's money and I will assist both sides."

The last penny had just been counted when the closing-gong struck.

"It balances exactly," reported Miss Overton, and then they all marched out while the orchestra played; but I fear they didn't keep step as well as usual They really couldn't, for their minds weren't on their feet, nor yet on the music, but on the great thing they had done for the school; and they were wondering if the Eighth Grade had done any better than they.

III

The next morning, when it was time for the arithmetic class. Miss Overton said:

"You were so much interested in working out your opponents' problems yesterday that I thought you might like to continue it to-day instead of having the regular arithmetic lesson. Miss Fowler has given me some statements regarding the amounts contributed by the other rooms in the building, and we will now find out what they contributed."

They prepared for this lesson with eagerness seldom shown for a regular lesson, although it must be acknowledged they were always good workers.

Miss Overton went to the blackboard and wrote the names of the grades in a column.

Teacher: Write form, filling in amounts as found.

Kindergarten..$ 1.12
First Grade..2.18
Second Grade..3.13
Third Grade.. \
Fourth Grade.. / 8.19
Fifth Grade..8.82
Sixth Grade..22.10
Seventh Grade...9.01
Eighth Grade..13.30
　　Total...$67.85

"As soon as we learn how much each room has contributed we will record it," said she.

"Then we can put our own down at once," remarked Annabel Jones, and it was done.

Miss Overton began:

"The Kindergarten children couldn't earn any money, and we didn't ask either them or the First Grade to help; but they heard their brothers and sisters talking about the plan, and of course they wanted to help, too. So their papas and mammas gave them pennies. In all, the Kindergarten brought one hundred twelve pennies. There are thirty-five children in the Kindergarten. How many pennies apiece did they average!"

Children, how many?

112 *pennies* ÷ 35 = 3 1/5 *times* . $0.03 1/5 *average.*

"In the First and Second grades the children were too young to earn much. However, seven in the First and nine in the Second Grade really earned their money by running errands and taking care of babies. In all, the First Grade brought two dollars and eighteen cents, and the Second three dollars and thirteen cents. The seven First Grade children earned forty-five cents, and the nine from the Second Grade, fifty-seven cents. How much money was given by the parents of children in these two grades?"

Can you tell?

$2.18 - $0.45 = $1.73 *given in First Grade.*

$3.13 - $0.57 = $2.56 *given in Second Grade.*

$4.29 *given.*

or

$2.18 + $3.13 = $5.31 *donated by both schools.*

$0.45 + $0.57 = $1.02 *earned by both schools.*

$5.31 - $1.02 = $4.29, *amount given by parents.*

How much did the two schools donate?

Answer: $5.31.

"There are thirty-seven children in the Third Grade and thirty-five in the Fourth. Eighteen of these brought five cents apiece. The rest of the children averaged thirteen and a half cents apiece. How much did the rest bring?"

37 *children* + 35 *children* = 72 *children in both grades.*

18 x $0.05 = $0.90 *brought by* 18 *children.*

72 *children* - 18 *children* = 54 *children who average* $0,135.

54 x $0.135 = $7.29, *amount brought by* 54 *children.*

How much did they all bring?

$7.29 + $0.90 = $8.19, *amount brought by Third and Fourth Grades.*

"In the Fifth Grade everybody earned money. For several weeks a contest had been on between the boys and girls, and on Experience Day it was found that the girls had brought four dollars and a half, while the boys had four dollars and thirty-two cents. There were sixteen boys and eighteen girls. Which group had brought the most per capita, and how much?"

How would you find this out?

$4.50 ÷ 18 = $0.25, *girls' average.*

$4.32 ÷ 16 = $0.27, *boys' average.*

$0.27 - $0.25 = $0.02 *boys brought more per capita.*

How much did the entire Fifth Grade bring?

$4.50 + $4.32 = $8.82 *brought by entire Fifth Grade.*

"In the Seventh Grade one boy brought forty-five cents; another fifty; one girl gave forty-two cents, and another thirty-five cents. The rest of the school together brought seven dollars and twenty-nine cents. How much did the Seventh Grade add to the fund?"

$0.45 + $0.50 + $0.42 + $0.35 + $7.29 = $9.01 *donated by Seventh Grade.*

"The Eighth Grade," continued Miss Overton, and one might have felt the tension in the room when she said "Eighth Grade," "averaged thirty-one and two-third cents with an enrollment of forty-two pupils. What was their contribution?"

The pupils worked as fast as they could.

42 x $0.31 2/3 = $13.30 *given by Eighth Grade.*

"Why, we 're ahead of the Eighth Grade, Miss Overton!" exclaimed Ralph, at which everybody applauded.

"But I can't understand how a Sixth Grade could beat an Eighth Grade," said Floyd, looking puzzled, when the applause had subsided.

"There are three reasons," explained their teacher. "The first is that we were well organized, and organization is essential to success. The second reason is that our competition, Hustlers versus Rustlers, kept up an interest. Thirdly, we had enthusiasm for a cause. These three factors were bound to make us win. There really is one more reason," added she. "The idea of earning money originated with us, and we naturally were more anxious than the other grades for its success."

Then they added up the column and there was still another chance for applause when they saw that the total amount was —

How much?

Answer: $67.85.

"I think we'd better sing our school song now," somebody suggested; and they sang it with spirit.

A little later Miss Fowler entered, and seeing the list on the board, said to Miss Overton:

"The teachers have together donated five dollars, so add that to your amount."

$67.85 +$5.00 =$72.85, *total.*

"How much does a radiopticon cost?" inquired Effie Wheaton.

"Fifty dollars will get us a good one," was the reply; and Dennis Ryan at once announced that after buying a radiopticon there would be a balance of—

How much?

$72.85 $50.00 = $22.85, *balance.*

"Why could not we give a radiopticon entertainment and sell tickets to our fathers and mothers? We did that in Chicago," said Flora Aldrich. "Then maybe what we earned with what we have left would buy a victrola."

"I think we might do that. I will talk it over with the teachers after school," promised Miss Fowler.

IV

The radiopticon was bought and the pupils planned to have their entertainment three weeks later.

When the Eighth Grade heard the result of the competition they were disgusted, and Josephine Ray, the most popular girl in the school, exclaimed:

"The idea of our letting a little Sixth Grade beat us like that! I will tell you what! We will ask Miss Fowler if we may not have a candy sale at the radiopticon entertainment. The boys will help us by giving money for the material and I 'm sure, our teacher will help us make the candy in the domestic science room some night after school, and we can each make a batch at home, besides. We can make fudge and pinocha and divinity and peanut brittle, just as we did last year at our candy-sale."

So they appointed committees and made their plans.

When the night of the entertainment arrived a committee of Eighth-Grade girls stood behind tables on each side of the assembly-room door and sold the candy they had made. It had been put up in boxes and paper bags, each in two sizes. Before the entertainment began they had disposed of the last package and could have sold more if they had had it.

They had sold seven boxes at twenty-five cents, eleven boxes at twenty cents, thirty-nine bags at ten cents, and thirteen bags at fifteen cents, which in all amounted to — How much!

$7 \times \$0.25 = \$1.75$ *received for 7 boxes.*
$11 \times \$0.20 = \$2.20$ *received for 11 boxes.*
$39 \times \$0.10 = \$3.90$ *received for 39 bags.*
$13 \times \$0.15 = \$1.95$ *received for 13 bags.*
$\$9.80$ *received for all.*

"Who is ahead now!" eagerly asked somebody, when the proceeds had been counted. Who was?

$\$13.30 + \$9.80 = \$23.10$, *total from Eighth Grade.*

"We are," announced Josephine Ray, "just exactly one dollar ahead, and we have —

How much for the victrola fund?

$\$22.85 + \$9.80 = \$32.65$ *for the victrola.*

In the assembly room the school orchestra gave a short programme and then several sets of pictures were exhibited by means of the radiopticon, while the principal explained them. The interested audience was taken on a trip through the Yellowstone, made a visit to Japan, and sailed through the Panama Canal. In short, they enjoyed three "personally conducted" tours, all for the small sum of ten cents. Besides these, the pictures of "Red Riding-hood" and "Mother Goose" were presented, to show how the radiopticon could be used for the lower grades. Last of all, a series of "Just for Fun" pictures was given, which the grown people enjoyed every bit as much as the children.

At the door seven dollars and thirty cents had been taken in, besides the tickets. The children had sold sixteen dollars and twenty cents' worth of tickets; so the entire entertainment brought in —

How much?

$\$16.20 + \$7.30 = \$23.50$, *proceeds of entertainment.*

If all who bought tickets attended, how many people were at the enter-tainment?

$23.50 x $0.10 = 235 *times.* 235 *people.*

The amount made added to what they already had, brought the victrola fund up to — How much?

$32.65 + $23.50 = $56.15, *victrola fund.*

"I'm glad the Eighth Grade had that candy sale, even if it did put them ahead of us," said Ralph to the Aldrich twins on the way home. "It gave us so much more money." A little later he added: "I told you our school could get anything that any other school could get! We have our radiopticon, and I bet we shall have a victrola before Easter,"

And, sure enough, they had!

Their Cost

SUGGESTIONS

Let each child make a booklet, as for "How Everybody Helped," giving it the title, "My Cost." Later, "Ralph's Cost" and "Dorothy's Cost" may be added.

Study the arrangement of problems before deciding on the size of paper to be used. System in arrangement is most important.

THEIR COST

Ralph and the Aldrich twins were on their way to school and Ralph had been telling Flora and Floyd some of his experiences on the farm.

"I have never been so surprised in my life as I was when I reckoned up how much I had eaten while I was at Grandpa's," he said. "You have no idea, if you have never stopped to count up, just how many meals you really do eat," he added. "Why, that would be something interesting to work out in the arithmetic class some day, wouldn't it? Let 's ask Miss Overton if we may do it."

"I think that would make a most interesting problem, and we certainly must try it," Miss Overton agreed, when, later in the day, Ralph told her of his idea; and she smilingly added: "Ralph, you seem to have a good deal of initia-tive."

"What 's that, Miss Overton?" asked Ralph.

"Look in the dictionary and see," she responded, and smiled again.

"That's just the way Uncle Ben and Grandpa do — let me find out things for myself, "said Ralph. "I like to do it. It makes me feel —

"Like a conqueror," suggested the teacher, as he hesitated for the right word.

"That's just it!" he cried.

So, in the arithmetic class the next day, instead of having the class do the examples on Page 210, Miss Overton asked:

"Children, how many meals do you eat a day?"

Of course everybody said three, and the children smiled because they considered their teacher's question a joke.

"How many days are there in a year!" she asked them next.

Most of them knew there were —

How many?

Year = 365 *da.*

"But next year, nineteen hundred sixteen, there will be three hundred sixty-six," ventured Chris Altman.

"That's true, Chris, "responded the teacher. "But that happens only once in every four years, you know."

Teacher: Discuss if necessary.

"We'll just say three hundred sixty-five," she went on. "Suppose you all find out how many meals you eat in a year."

How many do you eat in a year, children!

365 x 3 *meals* = 1,095 *meals eaten in a year.*

"How old were you when you began to eat three meals a day!" was the teacher's next question.

Several different answers were given to this, varying from one year to three.

"Suppose we say two years," said Miss Overton. "That will do for an average. Now, for how many years have you eaten three meals a day?" she asked.

Naturally the answers were different in different cases. Those who were Ralph's age found that this interesting process of eating three meals a day had been going on steadily for nine years.

"Each one may now find out how many meals he has eaten in his entire life," Miss Overton said. "Of course the answer will not be exact, but it will be what we call a close estimate."

How many meals had Ralph eaten?

9 x 1,095 *meals* = 9,855 *meals.*

"How much do you suppose a meal costs, on an average, in your home?" the teacher asked. "Ask your parents to-night to estimate it for you, and make a note of it. People live so differently that the cost will be quite different in different cases."

"After I got back from Grandpa's," observed Ralph, "I asked my mother, and she said in the way we lived she thought a meal would cost about nine cents on an average. But my mother is very economical," he added.

"Then you can find out now what your meals have cost," said his teacher.

How much had Ralph's meals cost?

9,855 x $0.09 =$886.95, *entire cost of meals.*

"It doesn't seem possible!" Ralph exclaimed, gazing at the figures.

"All of the rest of you children, remember to get your estimates to-night," said Miss Overton, "so that you can find what your meals have cost."

"Mayn't we find how much we've cost altogether?" asked Ralph. "I mean, how much our clothes and everything have cost, "he explained.

"How many would like to do that?" asked Miss Overton. The vote for it was almost unanimous.

"This, then, is our arithmetic lesson for to-morrow. "

Would you like to do this, too, children?

"We will begin with clothing," Miss Overton went on. "Write in a column the articles I name and leave a space after each item in which to write the cost. Get an estimate from your parents to-night and write it in the space."

"If your mothers haven't time to tell you all this, do the best you can by yourselves, and perhaps I can help you before school to-morrow," said Miss Overton; for she knew how very busy most mothers are.

Have the pupils write the items in the table as they are read; or the teacher may write them on the blackboard"

How many of each per year?	Average Cost
1 Suits (boys)	
Dresses (girls)	
2 Overcoats or sweaters (boys)	
Coats or sweaters (girls)	
3. Pairs shoes............................	
4 Pairs rubbers..........................	
5 Pairs stockings	
6 Undersuits	
7. Hats	
8 Caps	
9 Pairs mittens.........................	
10 Handkerchiefs	
11 Ties (boys)..........................	
12 Hair-ribbons (girls)	

II

Many of the children were so interested that they worked out their estimates at home. Several worked them, but forgot to leave out two years, so they had to work them again in school.

Let us work ours now, and later we will see how our expenses compare with Ralph's.

After each pupil has worked his own clothing problem either dictate Ralph's or put it on the hoard for copying. (See following page, and later in this chapter.) Keep the lists for future reference.

Let us see if we get the same results that Ralph did.

CLOTHING			PUPIL'S WORK	
Article	No. bought	Average cost	Total cost for 1 yr.	Entire life (less 2 yr.)
1 Suits	2	$3.75	$ 7.50	$ 67.50
2 Overcoats or sweaters.	1	3.50	3.50	31.50
3 Pairs shoes	3	2.00	6.00	54.00
4 Pairs rubbers	2	.50	1.00	9.00
5 Pairs stockings......	8	.15	1.20	10.80
6 Undersuits	4	.50	2.00	18.00
7. Hats	2	.85	1.70	15.30
8 Caps	1	.50	.50	4.50
9 Pairs mittens........	1	.45	.45	4.05
10 Handkerchiefs	8	.15	1.20	10.80
11 Ties	4	.15	.60	5.40
Total............	.....	....	$25.65	$230.85

When all of Miss Overton's boys and girls had found the cost of their food and clothing, the results were carefully put away for later use.

"What shall we estimate besides those two items?" was the next query.

"Sometimes we spend money on the moving picture shows." suggested Peter Lewis.

"Then we might estimate what we pay for amusements," responded his teacher. "Let us take this by the month instead of the year, and count not only the moving-picture shows but the circus, excursions, theaters, concerts, and sociables. Make a note of them. Now, what else?" she asked, as the children were silent.

"We pay out a good deal of money for candy and ice-cream cones, "said Mabel Gartney.

"Yes, and chewing-gum, popcorn, and soda-water, too," said Miss Overton. "Let us consider them under the head of 'Treats,' and estimate their cost by the month, also."

This, like the first items, was written in the notebooks.

"For what else do you spend money?" the teacher persisted.

"Sometimes we're sick and have the doctor or buy medicine," said Ema Lawton.

"Take that by the year and label it 'Illness,'" said Miss Overton, and they wrote it down.

"Are there any other items?" she asked.

"We travel sometimes," suggested Ralph, having in mind the money Uncle Ben had paid for his trip to the farm.

"Under 'Travel' let us count all kinds of rides that cost money — on trains, boats, and street-cars, and in automobiles and carriages — and estimate the cost by the year," advised the teacher.

"We buy books and pencils and paper for school use. Ought that to be counted?" asked Ernest Belden.

"Yes, indeed! It's a regular expense," was the reply. "That is another by-the-year item. However, let us take away five years instead of two, when we estimate that item. Make note of that as 'School Material.'

"Is that all?" Miss Overton asked, after a long pause. The children thought hard, but could not suggest another item.

"What did you do when we purchased our radiopticon?" asked the teacher.

"We gave money," was the answer.

"At what other times do you give money?"

"When we go to church and Sunday-school."

"Yes," said Miss Overton. "Do you give money at any other time?" she asked.

"At Christmas!" suggested one boy.

"At Thanksgiving, for dinners for poor people," said another.

"On Memorial Day, for decorating the graves of dead soldiers," one of the girls said.

"On birthdays!" cried some one else.

"Yes," replied Miss Overton. "We give money for a good many different things that benefit a community or make others happy; so we will enter their cost under 'Benevolences,' and estimate it by the month." She wrote the word on the board.

Teacher: Do the same.

"Don't you want to look up that word in your dictionaries immediately?" she continued.

They did so with alacrity.

Suppose you do the same, children.

"Bring all your estimates to-morrow and we will find out what we've cost, "Miss Overton said in conclusion.

III

The next day when the class was called the teacher arranged on the black-board the list given the previous day (in order to systematize the work) and the children copied it.

Teacher: Write and let the children copy.

	Per Mo.	Per Yr.	Entire Life (less 2 yr.)
1 Benevolences ..			
2 Amusements ..			
3 Treats			
4 Travel			
5 Illness........			
6 School material			(less 5 yr.)
Total.......			

They filled in the blanks and worked out their several problems. There was a great difference in the problems, to be sure, for in some homes there were more money and fewer children than in others. But all the pupils were surprised at the results.

"Now, let us put it all together and find our entire cost," suggested Miss Overton.

So they ruled and headed a paper like the one which the teacher had placed on the board.

Teacher: Rule this final one carefully, to be kept. After the children have finished their own estimates let them draw another form and dictate Ralph's.

MY COST

NAME.............................. AGE........			
	Per Mo.	Per Yr.	Entire Life (less 2 yr.)
1 Food			
2 Clothing......			
3 Benevolences ..			
4 Amusements ..			
5 Treats			
6 Travel			
7 Illness........			
8 School material	(less 5 yr.)		
Total.....			

Ralph's estimate was this:

RALPH MERTON'S COST   AGE 11

	Per Da.	Per Mo.	Per Yr.	Entire Life (less 2 yr.)
1 Food	$ .27		($ 98.55)*	($ 886.95)
2 Clothing			25.65	(230.85)
3 Benevolences ..		$ .32	(3.84)	(34.56)
4 Amusements ...		.25	(3.00)	(27.00)
5 Treats		.16	(1.92)	(17.28)
6 Travel			2.25	(20.25)
7 Illness			2.00	(18.00)
8 School material.	(less 5 yr.)		1.56	(9.36)
Total.......			($138.77)	($1,244.25)

*NOTE. All numbers enclosed in parentheses are for the teacher only. The others are to be dictated.

The estimate of Dorothy Perkins, whose father was a banker, was quite different from Ralph Merton's; for Dorothy's father had much more money to spend on his children than Mr. Merton could spend on Ralph. Her cost, the children found, was as follows:

DOROTHY PERKINS'S COST   AGE 13

		No. of	Cost of	Per Yr.	Entire Life (less 2 yr.)
1	Dresses	6	$8.50	($ 51.00)	($ 561.00)
2	Coats	2	7.75	(15.50)	(170.50)
3	Pr. shoes	4	3.50	(14.00)	(154.00)
4	Pr. rubbers.........	2	.60	(1.20)	(13.20)
5	Pr. stockings........	10	.35	(3.50)	(38.50)
6	Pr. gloves and mittens	4	.75	(3.00)	(33.00)
7	Undersuits	6	.85	(5.10)	(56.10)
8	Hats and caps.......	4	3.50	(14.00)	(154.00)
9	Handkerchiefs	12	.25	(3.00)	(33.00)
10	Hair-ribbons	8	.27	(2.16)	(23.76)
11	Other articles			7.00	(77.00)
12	Food, per day $0.65..			(237.25)	(2,609.75)
13	Benevolences, per mo. $0.45......			(5.40)	(59.40)
14	Amusements, per wk. $0.25.....			(13.00)	(143.00)
15	Treats, per wk. $0.45.			(23.40)	(257.40)
16	Travel			20.00	(220.00)
17	Illness			10.00	(110.00)
18	School mat'l (less 5 yr.)			5.00	(40.00)
	Total............			($433.51)	($4,753.61)

If this work becomes tedious, let one row of pupils find the cost of dresses; another that of coats, etc. Vary to keep their interest from flagging. Call attention to the fact that Dorothy's food costs more than Ralph's entire expenses. Contrast the amount of her "Benevolences," according to her means, with Ralph's, etc.

"Sometimes we pay more for things than they are really worth; and then we feel that we have been cheated," Miss Overton observed, "and sometimes we pay less than things are really worth and then we say we have a bargain. I hope your parents have not been cheated in you, but that you are all bargains — that is, that you are worth more to your parents than you have cost them. And remember that the money expended on you is only a small part of your cost. Think of all the work and care bestowed upon you! The cost of that can never be estimated."

Ralph looked very serious as he made his way home that afternoon. He was mentally comparing the amount he had earned that one week with the total amount that his parents had expended on him. It had seemed very large at the time, but now it seemed extremely small.

"If only I didn't eat so much!" he said. "I eat more than I wear."

He pulled his "cost paper" out of his pocket and studied it.

"My food costs more than all the other things put together," said he, after a careful scrutiny, "and I don't seem able to help it."

He entered the kitchen where his mother was baking the most appetizing cookies. He could smell them as soon as he stepped into the porch.

"I suppose I've eaten thousands of cookies," he remarked, "and never thought before that they cost real money."

"What in the world are you talking about, Ralph?" asked his mother; and without waiting for a reply she held out the pan, saying: "Here, take a nice hot one. They're the kind you like best." And in spite of the cost he didn't stop until he had eaten three cookies.

"Mother," said he, when the third one had vanished from sight, "do you know that I have cost you and Father one thousand, two hundred forty-four dollars and twenty-five cents?"

"My dear child, what are you talking about!" cried his mother. "How do you know how much you have cost us? You seem to have figured it out pretty accurately," she laughed, "even to the number of cents! If I had been estimating I certainly should have left off the twenty-five cents. "

"We did it in school, from the estimate you gave me last night," Ralph replied. "Don't you remember! Do you really think I'm worth twelve hundred dollars, Mother?" he asked, looking earnestly up at her.

"Well, my boy, I wouldn't sell you for that," said his mother, as she smiled down into his serious face; "no, nor for ten thousand times that!"

"Then I must be worth it!" cried Ralph joyfully.

"Of course it costs a good deal to bring up a boy,"

Mrs. Merton explained, "but Father and Mother are glad to work and earn the money to give him every comfort, and even luxury, they can afford. We don't expect children to pay for themselves in money. They pay in the comfort they are to us, especially when they are helpful and good-natured and considerate of those about them, like one boy I know," and she again smiled approvingly at him.

"I'm so glad you feel that way about it," said Ralph, "for it worried me at first. I wonder if you think I'm a bargain!" and he repeated Miss Overton's observations on the subject.

"Indeed I do, my boy — I think you're a real bargain," replied his mother.

"My, but it makes a fellow feel important to know that he's worth twelve hundred dollars! "said Ralph, drawing himself up to his full height. "May I have another cookie. Mother?"

"Yes, take all you want," said his mother, and he took three more, and went out to play ball with the boys.

How the City Helps

Suggestions

Make a blank-book for each pupil by fastening together twelve or fifteen sheets of paper. A pretty cover will make the book more interesting to the child who is to own it.

This book, unlike the preceding ones, is to have only the summaries copied into it, according to directions given later. Aim for greater rapidity in the work, while still insisting on neatness and accuracy.

Have the pupils write the following "Contents" on the first page, for ready reference:

How the City Helps

I

Christmas with all its joys had come and gone, and another vacation was over. When school reopened, several new pupils entered Ralph's class.

One of these was a boy with whom Ralph at once became very friendly. His name was Jarvis Brooks and so far his life had been spent on a farm. His father had now rented his farm and moved into the city, for the sole purpose of giving Jarvis a better education than would be possible in the little district school which the boy had attended np to this time. On their way home from school Jarvis and Ralph had most interesting discussions about crops and stock and other farm matters.

One afternoon Jarvis suddenly interrupted one of these "farm talks," so dear to Ralph 's heart, by asking:

"Ralph, when do we have to pay for the paper that our teacher gives us every day to write our lessons on?"

"We don't have to pay for it at all," responded Ralph. "They just give it to us."

"*Who* gives it to us? Nobody ever gave me paper, in the school out in the country. I always bought it myself."

"Well, we don't buy ours here." said Ralph. "The teachers have always given us pencils and paper and pens and books and everything. Sometimes we buy tablets and spellers and pencils, just because we wish to, but we really don't have to."

"Do the teachers pay for them?" asked Jarvis.

"I suppose so," said Ralph doubtfully and then added: "No, I don't believe they do, come to think about it; but I really don 't know who does. Let 's ask Miss Overton to-morrow."

So the next morning the two boys appeared before the teacher's desk and Ralph began:

"Miss Overton, you don't pay for all the paper and pencils and things you give us to use in school, do you?"

"Indeed not, Ralph!" laughed Miss Overton. "I shouldn't have any money left for myself if I did that."

"Then who does?" persisted Ralph.

"Everything we use in school is supplied by the Board of Education," replied the teacher.

"What's the Board of Education, Miss Overton? I have never understood about it."

The teacher explained how two men were elected from each ward and one from the entire city every two years by all the voters, to make up a committee, or "Board," whose duty it was to see that the schools were properly cared for.

Teacher: Explain local Board if it is constituted differently.

"They must have a lot of money," said Jarvis, "if they pay for everything that all the children use."

"Of course they do not use their own money," explained Miss Overton. "They use the taxes which are set apart for that purpose."

"Where do the taxes come from?" asked Ralph.

Do you know, children?

Discuss.

Miss Overton liked to answer questions of that sort when they were asked by boys who were really interested; so she explained:

"It costs money to make improvements. In order that the schools, streets, parks, and other things that are of general benefit to a community may be taken care of, each person has to pay a certain number of cents on every hundred-dollars' worth of property he owns. This money is all put together, and men are elected to decide how best to use it."

"If a man should say he had one hundred dollars' worth of property when he really had two hundred dollars' worth, he'd get out of paying some of his tax, wouldn't he? I suppose some men do cheat that way," remarked Jarvis.

"Perhaps some dishonest men try to do that, but it isn't very easy to do. A man called an assessor goes around to find out the value of the property and he decides on, or, as we say, fixes, the amount of the tax for each person," explained Miss Overton.

"Mightn't we do a few problems about taxes. Miss Overton?" asked Ralph, who had been listening intently. "I think we'd understand better if we worked them out ourselves. I know I always understand things better when I work them out myself."

"We might try a few simple problems, just to give you a definite idea of what taxes really are," his teacher replied. "When you reach the Eighth Grade you will learn all about them — possibly before that."

When the arithmetic class was called Miss Overton repeated what she had told Ralph and Jarvis before school, and added:

"All of you who would like to work a few simple problems in taxes may find out to-night the value of your fathers' property or of that of some one else in whom you are interested. Be sure to count all the property each person owns, just as the assessor would do."

The next day it was found that a number of the children had entirely forgotten to get the information Miss Overton had told them to, and several reported that their homes were rented and that their fathers had no property of their own. But Ralph and a few others were prepared for their lesson in taxes. They had written on slips of paper the amount of property they had found out about, with the owner's name, and had given these slips to Miss Overton.

Teacher: Ask the pupils to do this; or take the property valuation of some person in whom the children are interested.

"Jarvis Brooks's father," said Miss Overton, looking at Jarvis's slip, "owns a quarter-section of land, which the assessor valued at eighty-five dollars an acre. The tax rate this year is fifteen cents on a hundred dollars. What is his property tax?"

"What's a quarter-section?" asked Floyd Aldrich, who had always lived in a city.

Children, do you know?

Every farm boy knows this, and Ralph and Jarvis answered together: "One hundred sixty acres!"

Then the children all found out, first, that the assessed value was —
What?

160 x $85.00 =$13,600.00, *assessed value of farm.*

How many hundred-dollars is this?

Teacher: Drill on short method of dividing by ten, one hundred, one thousand, etc.

$13,600.00 x 100 = 136 *hundred-dollars.*

136 x $0.15 =$20.40, *Mr. Brooks's tax.*

"In some years more money might be needed for improvements than were needed this year," explained Miss Overton. "Then Jarvis's father would pay higher taxes, because the number of cents on each hundred-dollars would be more, or, as we generally say, the *rate* would be higher. Or again, less money might be needed, in which case Mr. Brooks's tax would be less. The rate might be ten or twelve cents on a hundred dollars, or even much less."

Ralph Merton's father, we remember, had just finished paying for his home the previous summer. It was worth twenty-six hundred dollars, but the assessor valued it at twelve hundred.

Teacher: *Explain that the assessor's valuation is always much less than the real value.*

"The tax rate in the city is higher than that in the country," Miss Overton explained. "This year our city tax is two cents on every dollar."

What was the amount of Mr. Merton's tax?

1,200 x $0.02 = $24.00, *amount of Mr. Merton's tax.*

"Dorothy Perkins's father owns a store building which the assessor valued at twenty-one thousand three hundred dollars," continued Miss Overton, "and a house valued at seven thousand five hundred dollars. As he lives in the city his rate is the same as that of Ralph 's father. What is his tax?"

The children worked the problem quite readily and found that his property was valued at —

How much?

$21,300.00 +$7,500.00 = $28,800.00, *value of Mr. Perkins's property.*

His tax was how much?

28,800 x $0.02 =$576.00, *amount of Mr. Perkins's tax.*

"That seems a good deal of money to give away," remarked Ralph.

"But it isn't given away," said Miss Overton. "Mr. Perkins has a good deal of property which owes its value to the city. It also is protected by the city from fire, and made more valuable by all the city improvements for which the money collected in taxes is spent."

"I thought you said the taxes were given to the Board of Education to pay for the schools," said Ralph.

"Only a small part of the city tax is used for the schools — probably not more than one-fourth at the most," explained Miss Overton. "This proportion will be different each year, and different in different cities. It is decided each year what it shall be. Let us suppose that it is one-fourth of the entire tax. Now find out how much Mr. Brooks's school tax would amount to if the same proportion of the country tax was used for the schools. What was his entire tax?"

"Twenty dollars and forty cents," promptly answered Jarvis. "Then his school tax would be —

How much?

¼ x $20.40 = $5.10, *Mr. Brooks's school tax.*

"Five dollars and ten cents," Jarvis announced, and the other children agreed.

"What will Mr. Perkins's school tax be?" asked Miss Overton.

In a minute the children had found that it was —

How much?

¼ x $576.00 = $144.00, Mr. Perkins's school tax.

"Louise Farrel's father," said Miss Overton, glancing at the slip of paper which Louise had given her, "owns five houses in the city. Their real values are (write these numbers, children) six thousand five hundred dollars; two thousand dollars; twelve hundred dollars; seven hundred dollars, and six hundred fifty dollars. They were assessed at four-fifths of their value. Mr. Farrel also owns a farm assessed at nine thousand dollars, where the rate is one-half cent on a dollar. What is his entire tax?"

The boys and girls looked bewildered, but Miss Overton smiled encouragingly and said:

"It isn't in the least hard. You have done more difficult ones many times. What shall we do first?"

Do you know, children?

Most of them said, "Find the value of all his property." Would you?

"Find just the value of the city property," said Jarvis.

"Yes," said the teacher, "but why?"

Why, children?

"Because," said Jarvis, "the tax rate is different in the country; so you must find his farm tax separately."

Was he right?

So the class first found the value of the city property, which proved to be — How much?

$6,500.00 + $2,000.00 + $1,200.00 + $700.00 + $650.00 = $11,050.00, *the value of all Mr. Farrel's city property.*

"What next?" asked Miss Overton.

What would you do next, children?

"Find how much of that they really taxed," said Dennis Byan. "You said only four-fifths of it."

Was Dennis right? How much of it was really taxed?

⅘ x $11,050.00 = $8,840.00, assessed value of Mr. Farrel's city property.

When the children had found this, it was very easy; for they all knew the rate was two cents on a dollar, so that the city tax would be —

How much?

8,840 x $0.02 =$176.80, *Mr. Farrel's city tax.*

They knew then that his farm tax would be — How do you find out how much it would be?

9,000 x $0.005 =$45.00, *farm tax.*

Then his entire tax would be — How much?

$176.80 + $45.00 = $221.80, *Mr. Farrel's entire tax.*

"Now," said Miss Overton, "find his School tax for the city."

What was it?

¼ x $176.80 = $44.20, Mr. Farrel's city school tax.

"And now his school tax for the country," said the teacher.

Can you find it?

"We can't find that," said Jarvis, "because the rate isn't the same in the country."

"That is correct, Jarvis," responded Miss Overton, "and as I'm not going to tell you the rate, you cannot find it at all. This must suffice for the subject of taxes until you are in a higher grade. How many of you understand it as far as we have gone?" she asked.

Up went the hands of most of the children and the interested faces proved more to the teacher than did the hands. Miss Overton, like most other teachers, could read children's faces, and in that way could tell whether they understood or not.

Do you understand taxes as well as these children did! Let me look into your faces.

"I notice," said Floyd Aldrich, "that some fathers pay higher taxes than others, but that we children all get the same number of books and the same amount of paper and other things. How many dollars do you suppose is paid out for each one of us in a year. Miss Overton?"

"Are you willing to do a good many problems for the sake of finding the answer to what Floyd has asked, children?" asked Miss Overton.

"Yes, indeed!" and "Yes, Miss Overton!" came from the class.

"Very well, then, we will," responded she. "What should you like to consider first?"

"The paper and pencils made us think about it first, "said Jarvis.

"Then I think it would be well to consider writing supplies to-morrow," said Miss Overton.

II

That afternoon, when the children and nearly all of the teachers had gone home. Miss Overton spent an hour or two in Miss Fowler's office, jotting down items from a big book. It was the book in which the principal recorded all the school expenses, and now it was to be of more real value than ever before, for it was to help the boys and girls learn some of the facts that it is well for boys and girls to know.

"Make a list of all the kinds of writing material we use in school," said Miss Overton the next morning. "There are eleven different kinds."

Nobody could think out the entire eleven alone, but each pupil thought of something, so finally the list was complete. Miss Overton put before each item the amount used in a year, and after it the cost.

Teacher: Place on the blackboard all numbers except those in parentheses, which will he added later. The parenthesis enclosures in all of the problems following are for the teacher only.

ANNUAL COST OF WRITING SUPPLIES

194 rm. Manilla paper	@ $0.10½	($ 20.37)
78 rm. composition paper	@ .22	(17.16)
2 rm. congress cap paper	@ 1.10	(2.20)
6 rm. hectograph paper	@ .37	(2.22)
17 gr. writing pencils	@ 1.75	(29.75)
28 gr. pens	@ .35	(9.80)
2½ gr. penholders	@ 1.50	(3.75)
344 spelling tablets	@ .06	(20.64)
60 writing tablets	@ .03½	(2.10)
240 folios for writing paper	@ .01½	(3.60)
36 rm. examination paper	@ .47	(16.92)
11 gal. ink	@ 1.20	(13.20)
Total...........................		($141.71)

"That looks like a bill," remarked Jean Bryce.

"Suppose we make it into one," suggested Miss Overton. "Do you know any firm in this city that sells that sort of supplies?"

"Brown and Brewster," cried Hans Schmidt. "I know, because my father works there."

"And to whom would the bill be made out?" queried Miss Overton.

Do you know, children!

"To the Board of Education," replied Ralph.

Teacher: Get name of local dealer or some large city firm and formulate bill. Have different pupil place amount of each item. Verify result.

When the bill was completed on the board. Miss Overton gave her pupils a pleasant little surprise. She passed to each one a booklet which she had made by fastening composition paper together with tiny brass fasteners. The books had pretty covers made of colored paper, and there was a special place where each pupil was to write his or her name. You may be sure the children did this carefully, for they always appreciated the little extra attentions paid them by their teacher. Every child in the class said, "Thank you," when he received his book. Perhaps that was because Ralph received his first; Ralph always remembered to be polite, and it was easy for the others to follow his example.

Either pass books already made or let the pupils make them.

Do you set good examples to others?

Opportunity to impress various ways in which we may help others by our example.

Then the children numbered their pages, numbering only on one side of the paper, after which they turned to the first page and wrote the word, "Contents" at the top. After that, whenever they did a problem they wrote the name of it on this page.

At the top of Page 2 they wrote, "Bill for Writing Supplies," and below this they made a neat copy of the bill just worked out.

Have pupils do same.

"What next?" asked Miss Overton, when this was completed.

"May we take the drawing supplies next?" asked Dorothy Perkins, who liked drawing, and whose drawings were always chosen for exhibits.

"Yes," said Miss Overton, "and it would be well to combine it with hand-work material." So she wrote for the next day's heading, up high on the front board, "Annual Cost of Drawing and Handwork Supplies."

"See how long a list you can think of," said she.

"I have thought of ten different drawing materials, but I'm not sure that I know the right names of all of them," said Ralph, the next morning.

"Write your list in a column, just as we did the one we made yesterday, and we will all help to make it longer," said Miss Overton.

Ralph wrote his ten items under the heading, the other children thought of five or six more, and Miss Overton concluded the list. It was this (*Teacher: Place on blackboard problem shown on next page*):

When it was all written out it certainly did look very long indeed.

"It's the longest problem I have ever seen in my life!" exclaimed Jimmie O'Hagan. "It will take forever to work it!" and he settled back in his seat with such a woe-begone look that everybody, even Miss Overton, laughed.

"Oh, no!" the teacher said. 'Forever' is too big a word to use that way! Have you ever walked up over the hill along that long winding road to the very highest point?" she asked.

"Yes, Miss Overton, ever so many times," replied Jimmie, wondering what that had to do with this problem.

"How did you get there — by one long step or a good many short ones?" asked Miss Overton.

"By a good many short steps, of course," said Jimmie; and everybody laughed again at the very idea of Jimmie taking so long a step.

"It's a long walk. Did you go all the way without stopping?" continued Miss Overton. "And were you tired when you reached the top of the hill?"

"No," replied Jimmie, "I stopped a few times and rested a while, and when I got to the top I wasn't a bit tired."

"Let us do this problem in the same way that Jimmie took his walk — take a good many short steps and stop two or three times," said the teacher, and they all seized their pencils and began to take the "short steps."

What were the "short steps"?

"You need not copy all of this on your papers," said Miss Overton; "just get the cost of each item separately and set them all down in a column. When you have finished the first seven we'll transfer the cost of them to the board."

You may do that way, too, children.

Have first seven products placed on board.

"Now, add them and we have —

Annual Cost of Drawing and Handwork Supplies

Item		Rate	Amount	Subtotal
16 pkg. large drawing paper	@	$0.08	($1.28)	
18 pkg. small drawing paper	@	.07½	(1.35)	
7 pkg. gray drawing paper	@	.09	(.63)	
3 pkg. white drawing paper	@	.10	(.30)	
38 pkg. large colored drawing paper	@	.14	(5.32)	
29 pkg. cutting paper	@	.11	(3.19)	
4 pkg. squared paper	@	.12	(.48)	
				($12.55)
1½ rm. black paper	@	1.50	($2.25)	
3 pkg. stenciling cloth	@	.23	(.69)	
20 cards mounting board	@	.05	(1.00)	
5 tubes dye	@	.15	(.75)	
10 pkg. eyelets	@	.25	(2.50)	
9 yd. burlap	@	.18	(1.62)	
4 lb. raffia	@	.40	(1.60)	
5 lb. rug yarn	@	.86	(4.30)	
				(14.71)
3 yd. binder's cloth	@	.25	($0.75)	
1 gr. pans paints	(per doz.)	·.20	(2.40)	
28 pt. paste	@	.30	(8.40)	
9 lb. hammock cord	@	.39	(3.51)	
2 lb. knitting yarn	@	.19	(.38)	
8 chip boards	@	.03½	(.28)	
1 box colored crayon	@	.34	(.34)	
3 sheets transparent paper	@	.07½	(.22)	
10 pkg. jute board	@	.60	(6.00)	
				(22.28)
Total........................				($49.54)

How much?

"Twelve dollars and fifty-five cents," said the class.

"Take the next eight in the same way," directed Miss Overton, and it was soon done. The answer this time was —

What?

Fourteen dollars and seventy-one cents.

The children then added the cost of the last nine items, which they found to be —

How much?

Twenty-two dollars and twenty-eight cents.

"Now add the three sums together," said Miss Overton.

They did so and found the total to be —

What?

Forty-nine dollars and fifty-four cents; and behold! the longest problem that Jimmie O'Hagan had ever seen, and which he said would take "forever," was finished!

"I always divide up a long column that way when I add," said Miss Overton. "I find I am less likely to make mistakes."

"I'll remember that," said Jimmie.

"It will take too much time to copy this all in your books, so let us just keep the summaries. On the third page write this"; and she dictated {Teacher: Dictate):

ANNUAL COST OF SUPPLIES

1 Writing supplies..$141.71
2 Drawing and handwork supplies.....................................49.54
3 Janitor's supplies...(447.61)
4 Printed matter...(27.97)
5 Miscellaneous supplies...(115.94)
     Total...($782.77)

"As soon as we find the total amount of each, put it in its proper place. Place the two that you have already found."

Children, you may do the same.

ANNUAL COST OF SUPPLIES FOR JANITOR

108 t. coal	@ $3.55	($383.40)	
15 gal. engine oil	@ .20	(3.00)	
7 gal. cylinder oil	@ .35	(2.45)	
350 lb. sweeping compound			
(per hd.)	2.25	$(7.87\tfrac{1}{2})$	
16 boxes matches	@ .21	(3.36)	
3 cases towels	@ 6.15	(18.45)	
6 mops	@ .50	(3.00)	
12 brooms	@ .35	· (4.20)	
			($425.73½)
			($425.74)
2 floor brushes	@ 3.20	($6.40)	
3 scrubbing brushes	@ .15	(.45)	
25 sponges	@ .01	(.25)	
3 gal. toilet soap	@ 1.25	(3.75)	
¼ bbl. cleaning soap	@ 5.00	(1.25)	
12 gal. floor oil	@ .40	(4.80)	
2 cans Dutch cleanser	@ .21	(.42)	
¼ gal. disinfectant	@ 1.40	(.70)	
Other small supplies		(3.85)	
			(21.87)
Total.....................			($447.61)

III

When the children entered their room the following morning they found this problem on the blackboard (*Teacher: Previous to this reading place on board the problem shown in the table above*):

When class time came nearly all of them had completed the problem. They knew that the line drawn across meant a stopping-place, so they really made two problems of it. Did you, children?

A few mistakes had been made in multiplying, but when these were all corrected Miss Overton asked, "What is the sum of the first eight items?" and Dorothy Perkins answered.

What do you suppose she said?

"Four hundred twenty-five dollars and seventy-three and one-half cents," said Dorothy.

"That's the same as four hundred twenty-five dollars and seventy-four cents," volunteered Peter Lewis. "When there's half a cent over, the person who's selling generally takes it."

"That is the rule in business," said Miss Overton; so they all changed the seventy-three and a half cents to seventy-four.

Louise Farrel, being called upon for the second sum, answered —
What?

"Twenty-one dollars and eighty-seven cents," was her answer; and when asked for the entire amount she said —
What?

"Four hundred forty-seven dollars and sixty-one cents," was what she gave, and this was correct.

"We certainly ought to keep clean, with three gallons of soap and three cases of towels," remarked Hans Schmidt.

"Yet I sometimes see hands, especially in marble season, that look as though we had no soap nor towels nor water," remarked Miss Overton.

Every boy in the class slyly inspected his own hands, and several looked extremely self-conscious.

"Now write the total amount on Page Three," said the teacher. "You may write on the same page, after 'Annual Cost of Printed Matter,' twenty-seven dollars and ninety-seven cents. This means the cost of all your report cards and other school records."

"This is the last of the problems about supplies," said Miss Overton the following day, as she indicated two lists on the blackboard.

Teacher: Place on hoard, previous to lesson,

"The one marked 'B' is for the boys and 'G' is for the girls. When you work them do not write the problems out: just put your products in a column and find the sum. I think you can do this in ten minutes. Try."

Have pupils work problems in same way.

ANNUAL COST OF MISCELLANEOUS SUPPLIES

B

5 boxes rubber bands	@	$0.24	($1.20)
10 boxes book-mending tape	@	.30	(3.00)
1 box transparent book-mender	@	.85	(.85)
2 rm. large book covers	@	2.05	(4.10)
3 rm. small book covers	@	1.37	(4.11)
16 boxes crayon	@	.17½	(2.80)
3 staff-ruled tablets	@	.12	(.36)
2 doz. thumb tacks	@	.03	(.06)
25 figure cards	@	.01	(.25)
4 boxes brass fasteners	@	.17	(.68)
Total...........................			($17.41)

G

¼ bbl. clay	@	$ 1.80	($ 0.90)
1 load sand (for tables)	@	1.00	(1.00)
9 pointers	@	.06	(.54)
3 doz. alphabet cards	(each)	.01½	(.54)
3 hectographs	@	.95	(2.85)
1 pt. red ink	@	.50	(.50)
2 bottles hectograph ink ·	@	.25	(.50)
Kindergarten material		26.70	(26.70)
Manual training material		35.50	(35.50)
Domestic science material		29.50	(29.50)
Total...........................			($98.53)

Every one finished in the ten minutes given, and there were only a few mistakes made. The boys agreed that their sum was —

What?

Seventeen dollars and forty-one cents.

The girls had a much larger amount. What was it?

Ninety-eight dollars and fifty-three cents.

The sum of the two was what, then?

One hundred fifteen dollars and ninety-four cents. This was written on Page 3, and then the total cost of the supplies for a year was found. Leonard Munroe said it was —

How much?

Seven hundred eighty-two dollars and seventy-seven cents. Was he right? Were you? How many were? *(Hands.)*

"What shall we take up to-morrow?" asked Miss Overton, and after a brief discussion the children decided that they would like to find the cost of all the books in the building.

79

"I have decided to begin with the readers and take them by grades," announced Miss Overton at the beginning of the arithmetic recitation the following morning. "Miss Fowler tells me that most of our readers average twenty to a set, so count twenty to each set unless I direct otherwise. Please write the numbers as I dictate."

Children, you may write them as I read them. "For the First Grade there are seven sets," said Miss Overton.

FIRST-GRADE READERS

2 sets @	$0.29	per book	(40 × $0.29)	($11.60)
2 sets @	.35	per book	(40 × .35)	(14.00)
1 set @	.42	per book	(20 × .42)	(8.40)
1 set @	.34	per book	(20 × .34)	(6.80)
1 set @	.33	per book	(20 × .33)	(6.60)
Total........................				($47.40)

The class found that the total cost of these readers was —
How much?
Forty-seven dollars and forty cents.

"Turn to Page Four and copy what I write on the board," directed Miss Overton, and she wrote (*Teacher: Write*):

COST OF READERS

First-grade	$ 47.40
Second-grade	(76.00)
Third-grade	(80.60)
Fourth-grade	(58.20)
Fifth-grade	(46.40)
Sixth-grade	(49.00)
Total........................	($357.60)

"Write the cost of the First-grade readers, and add that of the others as we find it, "said she.

Let US do the same thing, children.

"The Second Grade uses more readers than any other grade," Miss Overton stated, and she wrote *(Teacher: Write)*:

SECOND-GRADE READERS

3 sets @	$0.32	per book	(60 × $0.32)	($19.20)
2 sets @	.42	per book	(40 × .42)	(16.80)
2 sets @	.33	per book	(40 × .33)	(13.20)
1 set @	.29	per book	(20 × .29)	(5.80)
1 set @	.27	per book	(20 × .27)	(5.40)
2 sets @	.39	per book	(40 × .39)	(15.60)
Total........................				($76.00)

A few children forgot that there were twenty books in a set.

Did you?

Of course the products of those who did were much too small, but they multiplied the product which they already had by twenty or forty or sixty, according to the number of sets given, and the problem came out right, after all.

Those of you who have made that mistake, try this.

The class found that the cost of the Second-grade readers was —

How much?

Seventy-six dollars, even; and the pupils all recorded the sum neatly on Page 4.

You may record it in your books, too.

Miss Overton then dictated the Third-grade reader supply {Dictate):

THIRD-GRADE READERS

3 sets @ $0.43 per book (60 × $0.43)	($25.80)
2 sets @ .45 per book (40 × .45)	(18.00)
2 sets @ .48 per book (40 × .48)	(19.20)
2 sets @ .44 per book (40 × .44)	(17.60)
Total..........................	($80.60)

Fay Leland had the answer first, and when the others had finished, Miss Overton asked her to give it, thinking that perhaps she had made a mistake because she had worked so fast, which often happens. But Fay was right.

What was her total?

Eighty dollars and sixty cents.

So Fay proved that children can work rapidly and accurately at the same time, if they just keep their minds concentrated.

This gave Miss Overton an idea.

"Do not work the next until I ask you to do so," said she. She then wrote these three problems on the board *(Teacher: Write)*:

FOURTH-GRADE READERS

3 sets @ $0.50 per book (60 × $0.50)	($30.00)
3 sets @ .47 per book (60 × .47)	(28.20)
Total..........................	($58.20)

2 sets @ $0.47 per book (40 × $0.47)	($18.80)
1 set @ .48 per book (20 × .48)	· (9.60)
2 sets @ .45 per book (40 × .45)	(18.00)
Total..........................	($46.40)

SIXTH-GRADE READERS (25 in set)

2 sets @ $0.48 per book (50 × $0.48)	($24.00)
2 sets @ .50 per book (50 × .50)	(25.00)
Total..........................	($49.00)

"This time," said Miss Overton, "work as rapidly as you can *without making mistakes*. Stand in the aisle as soon as you have finished. Who will be among the first ten to rise? Begin!" and they worked as fast as they could.

Let us try the same plan, children. . Jarvis, Ralph, and Flora were the first three to finish, so they were asked to give their answers after seven others had finished, and they were all three correct, which again proves that if children will think carefully they can work rapidly without making mistakes. Their answers were —

What for each grade!

Fifty-eight dollars and twenty cents for the Fourth Grade; forty-six dollars and forty cents for the Fifth Grade, and forty-nine dollars for the Sixth Grade. Then it was only the work of a moment to record the amounts on Page 4 and find the total cost of readers, which was —

How much, children?

Three hundred fifty-seven dollars and sixty cents.

V

The first thing the arithmetic class did the next day was to copy on Page 5 a list that Miss Overton had placed on the front board.

Teacher: Previously write this list.

COST OF OTHER TEXTBOOKS

Arithmetics	($ 60.52)
Spellers	(34.88)
Music Readers....................	(61.30)
Geographies	(83.52)
Hygienes	(79.57)
Geographical Readers	(74.35)
History Readers..................	(33.50)
Special books .....‚.............	(34.99)
Total	($462.63)

Then she wrote these two lists on the board, telling the boys to find the cost of the arithmetics and the girls that of the spellers (*Teacher: Write*):

ARITHMETICS				SPELLERS		
65 @	$0.28	($18.20)		52 @	$0.16	($ 8.32)
94 @	.32	(30.08)		112 @	.18	(20.16)
34 @	.36	(12.24)		32 @	.20	(6.40)
Total.........		($60.52)		Total.........		($34.88)

When she had given them time to finish, Miss Overton asked Dennis Ryan and Kate Caswell to complete the problems on the board, which they did. All the children agreed with their answers, so of course these must have been correct.

Appoint a boy and a girl to do the same.

I wonder if you will have the same answers that Elate and Dennis had. (*Verify.*)

Then the teacher wrote three lists on the board like those you now see. (*Lists should have been written previously, with names of books omitted.*)

MUSIC READERS			GEOGRAPHIES			HYGIENES		
36 @	$0.15	($ 5.40)	72 @	$0.60	($43.20)	35 @	$0.39	($13.65)
48 @	.21	(10.08)	36 @	1.12	(40.32)	36 @	.38	(13.68)
50 @	.25	(12.50)				38 @	.42	(15.96)
98 @	.34	(33.32)	Total...		($83.52)	37 @	.44	(16.28)
						40 @	.50	(20.00)
Total...		($61.30)				Total...		($79.57)

"I believe the boys can work out the cost of the music readers and geographies while the girls are finding the cost of the hygienes," said Miss Overton. "Suppose we have a race, all beginning on signal just as the boys do in the relay race. When I say, 'Get ready!' all take pencils; 'Get set!' writing position; 'Go!' all begin. Keep your minds steady and *don't make mistakes.* The instant you have finished, bring your papers to me."

She had purposely left off the names of the books, fearing that some eager boy or girl would begin before the signal was given. She now wrote the names Music Readers, Geographies, and Hygienes above the problems and gave the signals. When she said "Go!" you should have seen her pupils work!

Let us race in the same way.

Put names above lists and give signals, following plan here given.

The race lasted ten minutes and was extremely exciting. As each paper was handed in the teacher made a record of the number of minutes used, discarding all answers that were incorrect as "fouls." When all were checked up she added the number of minutes used on each side and found the average number per pupil.

How did she do this?

If it is desired, the pupils may work out their own averages.

It developed that the girls' average was seven and seven-eighths minutes and that of the boys seven and three-fourths.

Which side won? How did you find out?

7 ⅞ mm.-7 ¾ min. (=7 ⅞ mm. - 7 6/8 min.)= ⅛ min. the boys were ahead.

Then the boys applauded, and Dorothy Perkins remarked, "You oughtn't to applaud very much for such a little 'beat' as that — only one-eighth of a minute." But she smiled as she said it, so they knew that she took the girls' defeat good-naturedly, just as people ought always to take defeat. To be cross and disagreeable when one is beaten is very unsportsmanlike.

Opportunity for discussing fairness, etc.

"A beat's a beat, whether it's big or little," sagely remarked Ralph.

VI

"There are three more problems about books," said Miss Overton. "Are you tired of this kind of work?"

"No, indeed!" exclaimed most of the children, and then some one asked if they might have another race. Miss Overton put the question to a vote and the majority voted for it.

"Suppose this time we all race together, and the first ten correct ones shall be called the winners. All who have correct work, if not among the winners, shall receive special mention. We will observe the same signals, and hand in the work the instant it is finished, just as we did the last time."

Then Miss Overton raised a map that had been pulled down over a portion of the front board, and there the children saw their problem all ready for them.

Teacher: Have same on board.

At the same instant she gave the signal to begin.

Shall we race, too?

When the race was over and the papers all marked, Miss Overton wrote the word "Winners" high up on the board and the first four out of the ten names that she wrote under it were Jarvis, Ralph, Floyd, and Flora, and under "Special Mention" there were four names headed by Jimmie O'Hagan's.

GEOGRAPHICAL READERS

25	@ $0.45	($11.25)
30	@ .47	(14.10)
36	@ .49	(17.64)
28	@ .52	(14.56)
30	@ .56	(16.80)
Total.......		($74.35)

What answer was gotten by all these pupils?

Seventy-four dollars and thirty-five cents.

"There are just three sets of History Readers," said Miss Overton, when the excitement over the race had subsided. "Take the numbers as I read them."

You take them, too, children.

As soon as these were written the children

HISTORY READERS

32	@ $0.35	($11.20)
30	@ .37	(11.10)
28	@ .40	(11.20)
Total.......		($33.50)

found themselves working as fast as though they were still racing. You see, they were beginning to form a habit. (*Discuss habits.*)

In a very few minutes the class announced that the sum was —
What?

Thirty-three dollars and fifty cents.

"The Board of Education is supposed to supply free textbooks only through the Sixth Grade," said Miss Overton, "but there are often good reasons why some of the children above that grade cannot purchase their own books. Their mother may be a widow with little means, or their father may be ill, or if not ill he may have a large family and receive small wages; or their families may be unfortunate in some other way. Misfortunes may overtake any of us, however prosperous we may be to-day.

"In order that the children from such homes may have the same opportunity for an education as other children, the Board of Education supplies each school with special books, where these are needed. We have a number of special books in our building," said Miss Overton. "Their total cost is thirty-four dollars and ninety-nine cents. You may write this on Page Five and then find the cost of all the textbooks in the building except the readers."

They did so and found the total cost to be —
How much?

Four hundred sixty-three dollars and three cents.

VII

"You are to have only two more book problems," said Miss Overton the next morning, soon after school opened, "and here they are."

The children looked at the board and saw this (*Teacher: Have on board*):

"I am adding the cost of our maps, globes, and charts to that of our reference books because they are used for practically the same purpose," explained Miss Overton.

The class found the cost of the reference books to be —
What?

Two hundred seventeen dollars and forty cents. That of the maps, globes, and charts was —
How much?

One hundred sixty dollars, making a total of —
How much?

REFERENCE BOOKS

Dictionaries	$ 26.00
Encyclopedias	36.00
History reference books............	15.00
Geography reference books..........	19.00
Drawing reference books............	13.00
Miscellaneous reference books.......	12.00
Estimated value of circulating library.	96.40
Total.......................	($217.40)

MAPS, GLOBES, AND CHARTS

25 wall maps	@ $ 0.85	($ 21.25)
5 reading charts	@ 2.60	(13.00)
5 globes	@ 3.15	(15.75)
3 history charts	@ 4.00	(12.00)
1 set of relief maps	@ 98.00	(98.00)
Total......................		($160.00)

$217.40 + $160.00 = $377.40, *cost of references.*
Three hundred seventy-seven dollars and forty cents.
"Make a record of this amount on Page Six," directed Miss Overton.
Do the same, children.
"Then turn to Page Seven," she continued, "and copy what I write on the board." She wrote this (*Write*):

To be certain that they had all copied the right numbers Miss Overton called on Jarvis to read the amounts. He then read the entire sum, which was —

TOTAL COST OF BOOKS

Readers	$ 357.60
Other textbooks	462.63
References	377.40
Total.........................	($1,197.63)

What?
One thousand one hundred ninety-seven dollars and sixty-three cents.

"On an average," said the teacher, "a book lasts about ten years. How shall we find the annual cost of our books?"

After thinking a while they found the right way.

What was the right way?

"Divide the total cost of the books by ten," said Erna Lawton.

The class did this, but I regret to say that a few of the children set down the numbers and did an example in short division, and one pupil even did it by long division. When Miss Overton saw this she said to the boy who did it — I will not tell you his name for fear you will think him stupid, though he really isn't — "I'm actually ashamed of you!"

Then she had them all divide several numbers by ten, one hundred, and one thousand, the quick, easy way.

Can you? (*Impress this.*)

The average cost of books in the building they found to be —
How much?

One hundred nineteen dollars and seventy-six cents, after deducting the 8/10 .

"I think we ought to be very careful not to waste any of the material that the Board of Education furnishes us, when it costs so much money; don't you, Miss Overton?" said Gertrude Blake.

"I do indeed, Gertrude!" the teacher answered.

"And I think we children ought to be more careful of our books than we are," suggested Eva Moreland, who was one of the most careful pupils in the school.

"I wish Dennis had said that instead of Eva," remarked Miss Overton. Everybody in the room knew that Dennis had to pay fines on his books oftener than any one else.

"I always pay my fines," said Dennis, looking rather uncomfortable.

"But fines do not make the book clean and whole again," said the teacher. "They may help a little toward the price of a new book, but somebody has to use the soiled one."

"I have never thought of it in that way," Dennis confessed.

"How would it be to put what Gertrude and Eva said into a resolution and all sign it?" asked Miss Overton.

Of course they wanted to do it, after they had talked about the word "Resolution" and found that to "pass a resolution" is what grown people do when they all together want to promise to do something for the good of a community.

The next morning before school Miss Overton passed a sheet of examination paper on which the Resolutions were written, ready for the children to sign.

"Read it very carefully," she advised, "and do not sign it unless you really mean to do what it says."

The teacher may have some resolutions copied and passed, for signatures. Change to fit local conditions.

This is what the children read:

"**Whereas,** The Board of Education spends a great deal of money each year for supplies and books for our school, and

"*Whereas*, the boys and girls are given these supplies and books absolutely free, be it therefore

"*Resolved*, that we, the undersigned, do hereby agree not to waste any of these supplies; and be it further

"*Resolved,* that we keep our hands clean so that our books will not become soiled; that we do not turn down the comers of leaves; that we do not keep loose papers in our books; that we try to keep them from dropping on the floor, as doing so loosens the binding; that we always keep our books well covered, and, lastly, that we never mark them nor deface them in any way.

"Signed......................................"

Before they really understood it, Miss Overton had to read the paper aloud. She explained that "whereas" meant, here, "because," and that grownup people always used that form. Every boy and girl then signed it, and it really did a great deal of good, because it made them remember. Probably no child really means to waste material or spoil his books. He just forgets, and then we say he is careless. The right sort of community spirit is a great help to careless people. It makes them remember to do right.

VIII

When the children entered the room the following morning, some of them looked discouraged to see a new problem on the board.

Teacher: Have the problem (shown on the opposite page) on the board, covered, when the children enter the room. Before beginning the race see that $.00 ¾ is understood.

But Miss Overton smiled at them and said: "Remember, it's just like walking — one step at a time."

"What does that word after 'Handwork' mean?" asked Sigmund Wolski.

"What are dictionaries for?" queried the teacher, and at the hint out came the dictionaries from the children's desks. Several read their definitions aloud and all learned that the word "Appliances" means —

Do you know, children? If not, look it up.

"Let us form teams for working this problem," said Miss Overton. "Ralph and Flora were good captains before, so they may choose sides for this race. "

Appoint two captains and conduct a race.

COST OF DRAWING AND HANDWORK APPLIANCES

4 doz. paste brushes	@ $0.60		($ 2.40)	
4 doz. boxes colored crayon	(each)	.10	(4.80)	
4 doz. erasers	(each)	.04	(1.92)	
3 doz. compasses	(each)	.08	(2.88)	
3¼ doz. sloyd knives	(each)	.09	(3.78)	
3 doz. stick-printing sets	(each)	.08	(2.88)	
4 doz. drawing boards	(each)	.12	(5.76)	
3 doz. cardboard looms	@	.57	(1.71)	
4 doz. clay-modeling tools	(each)	.00¾	(.36)	
4 doz. stencil brushes	@	.60	(2.40)	
6 doz. foot rules	(each)	.02	(1.44)	
				($30.33)
3¼ doz. small wooden looms	(ea.) @	.18	($ 7.56)	
5 pkg. pack needles	@	.23	(1.15)	
4 pkg. tapestry needles	@	.07	(.28)	
1 box giant clips	@	.25	(.25)	
175 pr. scissors	@	.13	(22.75)	
40 paint brushes	@	.05	(2.00)	
2 gr. drawing pencils	@	2.50	(5.00)	
25 pieces pottery	(average)	.65	(16.25)	
2 paper cutters	@	3.00	(6.00)	
1 large rug loom	@	.75	(.75)	
1 punch and eyelet set	@	2.10	(2.10)	
1 whittling outfit	@	4.50	(4.50)	
				(68.59)
Total........................				($98.92)

The choosing was quickly done, and the children insisted on being called Hustlers and Rustlers as before. This race was conducted like the others, the children starting on signal and handing in their papers as soon as finished. The teams proved to be very evenly matched, the Hustlers winning in the first section and the Rustlers in the second, each by a very small margin.

The first section amounted to —

How much?

Thirty dollars and thirty-three cents. The second came to —

How much?

Sixty-eight dollars and fifty-nine cents.

Then what was the total?

Ninety-eight dollars and ninety-two cents.

Miss Overton then wrote this on the board (*Write*):

The children copied it on Page 8, and recorded the first amount.

You may do this now.

"The Hustlers and the Bus-

COST OF APPLIANCES AND EQUIPMENT

Drawing and handwork appliances.. $	98.92
Janitor's appliances..............	(143.25)
Domestic science equipment........	(658.00)
Manual training equipment........	(434.94)
Miscellaneous appliances	(128.63)
Orchestra equipment..............	(48.00)
Playground equipment............	(150.00)
Total.......................	($1,661.74)

tlers may have another race to-day," Miss Overton said then, "and this time I shall dictate the items instead of putting them on the board. Do not attempt to write the names of the articles; just take the numbers, indicating the multiplication like this"; and she wrote on the board:

$$275 \times 15 =$$

"If you do not have to multiply, just set down the number with your products. There are fifteen items. How would it do to make fifteen decimal points in a column, now, so that your numbers will all be ready for adding?" and she showed them at the board just what she meant.

Then she read the numbers and the children repeated them together after her as they wrote, to be sure they all had the right numbers with which to work. When all the items were given, she said 'Go!' and they went!

COST OF JANITOR'S APPLIANCES

275 ft. lawn hose	@ $0.15	($ 41.25)
10 whiskbrooms	@ .19	(1.90)
10 patent dusters	@ .85	(8.50)
1 radiator brush	@ .40	(.40)
1 lawn-mower	@ 6.50	(6.50)
3 wire door-mats	@ 1.50	(4.50)
3 cocoa door-mats	@ 1.50	(4.50)
1 wheelbarrow	@ .50	(.50)
1 hand-bell	@ 1.50	(1.50)
2 stepladders	@ 1.35	(2.70)
1 pencil sharpener	@ 3.25	(3.25)
1 eraser cleaner	@ 3.75	(3.75)
Fire-hose and chemicals	35.00	(35.00)
Pails and hampers	3.85	(3.85)
Tools, dustpans, etc.	25.15	(25.15)
Total..................................		($143.25)

This time Flora's team, the Rustlers, came out a minute and a half ahead, and the Hustlers took their defeat very good-naturedly. Then all the pupils put down the total amount on Page 8.

What was it!

One hundred forty-three dollars and twenty-five cents.

And that ended that day's lesson.

89

IX

"The next few problems are too simple for team work," Miss Overton said at the next arithmetic lesson. "There is no multiplying. Take the cost of the articles as I dictate. You need not even write the name of the article." (*Read.*)

DOMESTIC SCIENCE EQUIPMENT

Gas range..........................	$ 30.00
24 gas plates.......................	24.00
Refrigerator	16.00
Cupboard	75.00
Cooking tables......................	250.00
Stools	25.00
Cooking utensils	66.00
Dishes	52.00
Sewing machine.....................	35.00
Miscellaneous	85.00
Total.........................	($658.00)

They did this problem very quickly and every one had the correct answer. What was it?

Six hundred fifty-eight dollars.

Then Miss Overton gave the list of Manual Training Appliances, and it was such a short list that the children smiled when she said, "That is all." It contained just four items. (*Dictate.*)

MANUAL TRAINING APPLIANCES

25 work benches	@ $	9.95	($248.75)
1 locker	@	62.00	62.00
Tools		116.50	116.50
Extras		7.69	7.69
Total....................			($434.94)

They had been working such long problems, that this seemed "just like nothing," as Jimmie O'Hagan expressed it. However, the amount wasn't "just like nothing, "for it was almost four hundred thirty-five dollars — not a very small sum of money.

How much was it, to be exact?

Four hundred thirty-four dollars and ninety-four cents.

The problem about Miscellaneous Appliances was just a little longer, but very easy compared with most of the problems the class had done. In it there were seven items, which the children took down as Miss Overton read them.

Children, write them as I read them:

MISCELLANEOUS APPLIANCES

Kindergarten gifts	$ 90.00
Kindergarten blocks	15.00
Other Kindergarten appliances.......	7.85
Bird pictures	2.85
Printing outfit	1.75
Toy money	.90
Other appliances	10.28
Total........................	($128.63)

When it was completed the total was recorded on Page 8.

What was the total?

One hundred twenty-eight dollars and sixty-three cents.

"Ernest, you play in the orchestra, so you have a chance to observe the instruments," Miss Overton said. "Will you write on the board a list of all not owned by the children themselves? Put down all except the piano," she added.

So Ernest wrote (*Teacher: Write*):

Then she wrote the cost after each.

When she read his list over she said: "I'm sorry if we have the kind of drum Ernest has put in his list. I thought we had a *good* drum. A b-a-s-e man is a very bad man, so a b-a-s-e drum must be a very bad drum."

Base drum	$12.00
Snare drum	6.50
Orchestra bells	12.50
Symbol	5.00
Traps	1.75
Triangle	.75
Music racks........................	3.25

Miss Overton told him to add

Four sets music books................	$ 6.25

"The word is spelled wrong," said Ralph.

How should it be spelled?

"'B-a-s-s' is the right way to spell it," he added.

"Yes," Miss Overton replied. "And what is wrong with the word in the fourth line?" she asked. "Is that the correct way to spell the name of the instrument Ernest means?"

Is it, children? How should the word be spelled!

"It should be 'c-y-m-b-a-l,'" said Flora Aldrich; and Miss Overton said, "That is correct."

When Ernest had corrected the two misspelled words Miss Overton asked for the total cost of the instruments named in the list.

What was it?

"Forty-eight dollars," was the answer her pupils gave.

"Leonard, you may write the names of all the apparatus used on the playground, "said Miss Overton, and Leonard wrote these four items (*Teacher: Write*):

91

Giant stride ....
Horizontal bar ....
Slide ....;....... ....
Seesaw ....

"The first cost forty dollars; the second twenty; the third forty-five; and the last forty-five," Miss Overton stated, and by the time she had given the last price, the children could give her the total cost.

What was it?

One hundred fifty dollars.

When the cost of all appliances and equipment together was found it amounted to —

How much?

One thousand six hundred sixty-one dollars and seventy-four cents.

"Aren't we ready yet to find out how much a year's schooling costs each of us?" asked Floyd Aldrich the following morning.

"Not quite," answered Miss Overton. "What else is there besides supplies, books, and appliances?"

What else can you think of, children? I wonder if you will think of the same things that these children thought of.

"The desks and chairs and bookcases and — and waste-baskets," said Ralph.

"And the pictures and clocks," said Dennis.

"And the pianos," added some one else.

"Yes, there are still the furnishings and decorations to be considered," said Miss Overton. "I'll just tell you their cost by groups, as it would take too long to work out the cost of each." Then she dictated this from her little memorandum book:

Children, write as I read.

"What is the total, children? "asked she.

Children, give the sum.

"Two thousand seven hundred sixty-nine dollars and five cents," was the answer.

"The amounts are growing bigger and bigger," said Ralph. "Why, that's more than our house and lot together are worth — almost two hundred dollars more!"

COST OF SCHOOL FURNISHINGS

Teachers' and children's desks.......	$1,341.00
Chairs and bookcases..............	416.75
Tables (sand, sewing, kindergarten, and others).....................	90.75
Pianos, clocks, and window shades...	867.65
Other furnishings	52.90
Total..........................	($2,769.05)

Miss Overton then told them that the pictures, casts, friezes, and other decorations had cost five hundred sixty-nine dollars.

On Page 9 the children wrote this:

"Look up the word 'permanent,'" said Miss Overton, and

COST OF PERMANENT SCHOOL PROPERTY

Appliances	$1,661.74
Furnishings	2,769.05
Decorations	569.00

they did so, finding that it meant —

What, children? Look in your dictionaries.

"My two farmer boys may go out and count the bushes on the grounds," said Miss Overton, and she smiled at Ralph and Jarvis as she spoke. "Floyd and Jimmie may count the trees belonging to the school grounds."

Out they went in a hurry. They were so very much interested in what they were going to do that they forgot to be quiet. Miss Fowler, who was just about to enter the room as they left it, didn't say a word about the noise, for she saw by their faces that they were bent on an errand of importance, and she knew that they were noisy because they were in such a hurry to do this important thing. So she just smiled at them.

In a few minutes they returned, almost out of breath, so fast had they hurried.

"There are one hundred twenty-five of those prickly bushes in the hedge between the playgrounds and the lawn," Ralph reported, and Miss Overton said,

"One hundred twenty-five barberry shrubs."

"Twenty-six bushes in the northeast corner," reported Jarvis.

"Thirty of the same kind in the northwest corner," said Ralph.

"Fourteen on each side of the front walk," Jarvis added.

"And thirty-eight more in other places," Ralph concluded. "That makes in all —

"Let the class find out," interrupted Miss Overton. Then she repeated the numbers, omitting the barberry shrubs. "Twenty-six, thirty, fourteen on each side, thirty-eight. These are lilacs, snowballs, weigelia, honeysuckle, syringa, and rose-bushes," she explained. "How many are there in all?"

How many were there?

"One hundred twenty-two!" answered the class.

Floyd and Jimmie reported thirty-six trees around the outside of the cement walk, thirty-four inside the fence, and thirty more scattered about where they made the yard shady and cool on hot days.

How many trees were there in all?

"Just an even hundred!" exclaimed Dennis. Miss Fowler, who had remained in the room, had listened with interest to the report.

"Ralph," she said, "go to the office, please, and call Mr. Holden the nurseryman on the telephone. Ask him how much he would charge to furnish and set out one hundred twenty-five barberry shrubs, one hundred twenty-two mixed shrubs (like lilacs and snowballs) and one hundred elm trees. Tell him I asked you to find out," added she, for she knew that sometimes busy men did not take much time to answer boys over the telephone.

Ralph went and soon returned, saying that the barberry shrubs would cost twenty cents each, the other shrubs thirty-five cents each, and the trees about ninety-five cents apiece.

"Now find out how ranch they all cost," said Miss Overton, and they worked faster than ever, for they wished to show Miss Fowler just what good workers they were. Next to Miss Overton they liked Miss Fowler better than any one else in school, and they were anxious to win her approval

The hands went up promptly, and Mabel Gartney, being called upon, said:

"They cost altogether —

How much, children?

"They cost one hundred sixty-two dollars and seventy cents," said Mabel, and the children set this down on Page 9, under "Decorations ": (*Write on board*):

Trees and shrubbery..$162.70

"If they were worth that much when they were set out, how much do you suppose they are worth now that the trees give us such fine shade and the shrubbery such beautiful blossoms?" asked Miss Fowler, and Jimmie impulsively exclaimed:

"A hundred times as much!"

"I believe they really are, Jimmie," responded she, "but since you are working with costs instead of values, you'll have to use the number just found."

"Our building isn't so very new, but we have bigger and prettier grounds than almost any school I have ever seen," volunteered Flora Aldrich. "When I first came here from Chicago I thought this school wasn't very interesting, but I've changed my mind," acknowledged she, at which confession the other children smiled contentedly.

"It is that feeling which helps make a school interesting," said Miss Overton. "We call it school loyalty, and it is to the school just what the soldier's loyalty is to his country. Now open your books to Page Nine again and write these items under * Permanent School Property. "' She wrote on the board (*Write*):

Building	$40,000.00
Grounds	9,000.00
Total	($54,162.49)

"Forty thousand dollars for this building!" exclaimed Jimmie. "Isn't that a mistake!"

"That's what the record book says, so it must be true," said Miss Overton. "You didn't know that you were spending a good part of ten months each year in a forty-thousand-dollar house, did you?"

But Jimmie still looked incredulous.

"I've found the sum of all the permanent school property, "said Dennis. "It is —

How much, children?

"Fifty-four thousand one hundred sixty-two dollars and forty-nine cents. My, that's a heap of money! "

"Yes, that is a *great deal* of money, "said Miss Overton, just as a hint that "heap of money" is very poor form. "But we are not through yet, "she went on. "What other things does the city, through the Board of Education, pay for?"

"They have to pay the teachers," suggested Margaret Whitten.

94

"And the janitor," added Ernest.

"Yes, the teachers and the janitor of our building together receive eight thousand eight hundred sixty-five dollars per year. Write that number down to be used later. Are there any other teachers besides the ones who come here every day?"

"Yes, the drawing teacher," one child suggested.

"And the writing teacher," another said.

"And the music teacher," added a third.

"Let us make a list of those outside the building who come here to teach, or to help us in any way. With whom shall we begin?" asked the teacher.

"The superintendent," suggested Chris Altman.

So they made a list, in which Miss Overton inserted the names of officials, some about whom the children did not know. After the list was made she grouped them and gave the combined salary of each group like this (*Write on board list given below*):

Adding up, the class found the salaries of these officials amounted to how much, then?

Twelve thousand two hundred five dollars.

After explaining the duties of the Clerk of the Board, the stenographers, and the head janitor. Miss Overton asked:

"How many buildings do these people serve?"

The children counted them up and answered: "Nine."

Superintendent Clerk of the Board Stenographer `	..............$ 4,650.00
Writing Supervisor Drawing Supervisor Music Supervisor	.... 3,175.00
Domestic Science Teacher Manual Training Teacher	.... 2,150.00
School Nurse Truant Officer Head Janitor	................:...... 2,230.00
Total.......................	($12,205.00)

"Yes," said Miss Overton. "We are fortunate in being a small city, for we get a greater share of the time of these officials than we should if there were more schools for them to serve. How shall we find our share?" she asked.

"We must divide by nine," suggested Carl Cummings. This the children did, getting as an answer —

What?

$12,205.00 ÷ 9 = 1,356.11 *times*. $1,356.11, *our share*.

"That is not exact, but it is a good estimate," said Miss Overton.

They then added this amount to the eight thousand eight hundred sixty-five dollars annually received by the regular teachers and the janitor of the building, and found that the sum was what, then?

Ten thousand two hundred twenty-one dollars and eleven cents.

"That is practically what the city pays every year in salaries for the education of the boys and girls of just our building alone," said Miss Overton. "Turn to Page Ten," she directed.

The class did so and arranged the page like this (*Write on board*):

ANNUAL RUNNING EXPENSES

Salaries in building...............	$ 8,865.00
Special salaries ÷ 9...............	1,356.11
Other running expenses..........	(774.63)
Total.......................	($10,995.74)

"What does 'running expenses' mean?" asked Chris.

Do you know, children?

"I know," said Jarvis. "Running expenses are what it costs to keep the work going on."

Then the class named different running expenses, and with suggestions from Miss Overton, made this list (*Place on hoard as children name*):

Lighting	$ 24.50
Gas for cooking....................	28.47
Insurance	92.66
Telephone	20.00
Water rent........................	35.00
Repairs and improvements..........	495.00
Material for manual training........	44.00
Material for domestic science........	35.00
Total........................	($774.63)

When the annual cost of each was given, what was the total found to be?

Seven hundred seventy-four dollars and sixty-three cents.

This they added to their list on Page Ten, and then found that the entire running expenses for the year amounted to—

How much?

Ten thousand nine hundred ninety-five dollars and seventy-four cents.

"We didn't count the heating," remarked Jarvis.

"No, because we put the coal on the janitor's supply list," said Miss Overton. "We might have put it here if we had thought of doing so."

"Have you found the entire value of the permanent property?" asked Miss Fowler, who was again observing the work.

"Yes, we have it on Page Nine. It is fifty-four thousand one hundred sixty-two dollars and forty-nine cents," Miss Overton replied.

"Suppose," said the principal to the class, "that the city had not paid out that money for building and furnishing the school; what might it have done with it?"

"Rented it!" said Sigmund Wolski.

"Put it in the bank," said another boy.

"Yes, the city probably would have loaned it out. How much do you think it could get for the use of one dollar for a year?"

Discuss interest rates.

After a short discussion the children thought they might safely count on six cents for each dollar loaned.

"Just drop off the forty-nine cents from the total value of permanent property and see if you can find how much money the Board of Education could have gotten for the use of its money," said Miss Fowler.

The children thought hard for a few minutes, and most of them found a way to do it.

Can you do it? How?

They were surprised that it should be three thousand two hundred forty-nine dollars and seventy-two cents, for six cents seems so small; but Miss Fowler told them that that was about the fastest and easiest way there was of making money — to let money make more money for you. She then advised the boys and girls to begin saving while they were young, and start a savings account which should be earning money for them.

"It is much better than spending so much for moving-picture shows and ice-cream cones and gum," said she.

"What do you call the money that is paid for the use of other money?" Miss Overton asked the class.

What is it called?

"Interest," answered several of the children at once.

"Yes, you have done a problem in interest," said their teacher. "You have found the interest on fifty-four thousand one hundred sixty-two dollars for one year at six per cent. In the higher grades you will do a great many problems in interest, but of course they will be much harder."

"Since the city gives us the use of property which would bring them that much money each year, it is just the same as though it gave us the money to use, isn't it?" asked Miss Fowler, and the children all agreed that this was true.

X

The next day Miss Overton said: "We have worked a long, long time to find out how much money the city really pays out for each one of us every year, but I believe we can really find out to-day. Turn to Page Eleven in your books and at the top of the page write out that little problem in interest that you did for Miss Fowler yesterday."

Children, do the same.

"Now write this on the same page," she went on, and wrote on the board the following list (*Write*):

VIRTUAL AMOUNT EXPENDED ANNUALLY

Page 3	For supplies ....($	782.77)
Page 7	For books..............	(119.76)
Page 10	For running expenses...	(10,995.11)
Page 11	Interest on property....	(3,249.72)
	Total................	($15,147.36)

The children copied the form and looked up the amounts to put down by referring to pages 3, 7, 10, and 11. Then they added them all together and found that the amount for which they had so long been working was—

How much, children?

Fifteen thousand one hundred forty-seven dollars and thirty-six cents.

"There are three hundred thirty-two pupils enrolled in our building. How can you find how much money is paid out per pupil each year?"

Do you know? How?

"Divide the amount by the number of pupils," a number of the children answered. Then they all did it.

"Leave your problem all written out. We haven't had one of this kind for a long time," said Miss Overton.

Children, work it and leave all of your work.

When most of the class had completed it one of the pupils placed it on the board and those who had made mistakes found them.

"Long division is about the hardest thing in arithmetic," said Hans Schmidt, "because you have to divide and multiply and subtract, all in one problem."

"Yes," added Ralph, "and you have to add, too, to see whether or not you have subtracted right."

"That's why we should do a good many such problems," said Miss Overton. "They give

$$45.62\,(\tfrac{152}{332} \text{ less than } \tfrac{1}{2})$$
$$332)\overline{15147.36}$$
$$1328$$
$$\overline{1867}$$
$$1660$$
$$\overline{2073}$$
$$1992$$
$$\overline{816}$$
$$664$$
$$\overline{152}$$

us practice in all the important operations. Now, look again at your answers. You have forty-five dollars and sixty-two cents, which means that that is the amount virtually paid out for each pupil in this building each year."

"That's one dollar and forty-two cents more than my father's school tax," said Louise Farrel, "so he doesn't quite pay for my schooling for a year."

"But my father pays a good deal more than enough for me," said Dorothy Perkins. "His school tax is one hundred forty-four dollars," and she looked very well satisfied.

"How many children has your father in school, Dorothy!" asked Miss Overton.

"Four besides me," replied Dorothy.

"Then how much does the city pay for all five of you, if it pays forty-five dollars and sixty-two cents for each!" continued Miss Overton.

When Dorothy had worked this out, she looked rather crestfallen. "It amounts to two hundred twenty-eight dollars and ten cents," replied she.

"Then how much does the city really pay for the education of your family?" asked her teacher.

Find out, children.

$228.10 - $144.00 = $84.10, amount paid by city.

"I think," said Miss Overton, "that in the majority of cases children receive more from the schools than their fathers pay in through their school tax. And we know that there are many, many children whose fathers pay no school tax, but who receive just as much as the others. Why does a city educate those children?

Discuss before further reading.

"Why should a city wish to educate the boys and girls?" queried Miss Overton.

"So that the boys and girls can do more for the city when they grow up," answered Joe Beverley.

"That is one important reason," said Miss Overton. "It wishes to make good citizens of them, and an educated person is of much more value to a community than an uneducated one. Not only can he do more for a community, but he is worth more to himself. He can think greater thoughts and that makes life mean more. Usually he can earn more money and therefore dress better and have a better home and more amusements. As a rule he enjoys life more."

"Sometimes we boys get tired of school and want to stop," said Dennis Ryan. "It's lucky for us that our parents won't let us, isn't it?"

"The truant officer wouldn't if your parents would," remarked Sigmund Wolski, and they all laughed, for they knew that he had once had an experience with the truant officer. "Now," he added, "I'm glad he made me come."

"Well, this is our last lesson about how the city helps. Let us sum upi What have we learned from this long series of problems!" said Miss Overton.

"To multiply and add faster, "said Joe Beverley.

"About taxes," said Jerome Sanders.

"What a quarter-section is," said Walter Brown.

"That money can earn money, "said Erna Lawton.

"How to work interest problems," said Carl Cummings.

"I've learned how to spell 'bass' and 'cymbal'," said Ernest Belden, "and that 'appliances' are things to work with."

"I've learned from our races that problems don't count at all if the answers are wrong, so we ought to be very careful," said Sallie MacKay.

"I've learned we get out of the school more than our fathers pay in for taxes," said Dorothy Perkins.

"I've learned what the Board of Education is," said Leonard Munroe.

"And what an assessor is," said Effie Wheaton.

"I've learned that the man who is selling generally keeps the half -cent left over for himself," said Hans Schmidt.

"I've learned that a man's school tax is only a small part of all his taxes," said David Winchester.

"I've learned that it takes an awful lot of supplies and appliances to teach drawing and handwork," said Peter Lewis, and Miss Overton said reprovingly: "Yes, a *great many.*"

"I've learned that if you 're beaten in a race or a game you shouldn't be cross about it," said Jean Bryce.

"I've learned that we can save some of the school money by being more careful of the books and supplies," said Fay Leland.

"One thing I have learned is that the trees and bushes in the city cost money, so that we have no right to break off branches here. It never mattered on the farm," said Jarvis Brooks.

"I have never before known it took so many kinds of things to keep a school going; so I've learned that," said Eva Moreland.

"I've learned that you can make a long problem easy by splitting it up into two or three problems," said Jimmie O'Hagan.

"The best thing I've learned," said Ralph, "is that the city pays out all this money for us children so as to make good citizens of us, and I've made up my mind that I 'm going to be one when I grow up. My father's school tax is so little that it's just as though the city gave me that forty-five dollars each year. I reckoned it up, and if I go through the Eighth Grade I'll owe the city almost four hundred dollars. If I 'm a good citizen, won't that be one way of paying the city back, Miss Overton?"

And Miss Overton responded heartily:

"It's the best possible way, Ralph, and I hope you will always keep that resolution!"

The word "resolution" reminded Jimmie O'Hagan of something.

"Can't you write that up into a resolution and let us all sign it as we did the other one?" asked he.

So Miss Overton wrote on another big sheet of paper:

"**Whereas**, the city is each year paying out large sums of money to carry on the schools so that boys and girls may become educated and so be better and happier and able to do more to help others, therefore be it

"*Resolved*, that we the undersigned hereby promise to be as good as we can, and learn all we can while in school, so that we may all be good citizens when we grow up, and in this way pay back the city for the money it has spent for our benefit.

"Signed..."

Very seriously, every one in the class signed it.

"If all the boys and girls in our country should make that resolution and never break it, how many good citizens there would be when they all grew up!" said Miss Overton.

I wish they all would, don't you?

www.ingramcontent.com/pod-product-compliance
Lightning Source LLC
Chambersburg PA
CBHW021936040426
42448CB00008B/1094